THE LIVING EARTH

QUAKES, ERUPTIONS, AND OTHER GEOLOGIC CATACLYSMS

REVEALING THE EARTH'S HAZARDS

REVISED EDITION

JON ERICKSON

FOREWORD BY ALEXANDER E. GATES, PH.D

Facts On File, Inc.

QUAKES, ERUPTIONS, AND OTHER GEOLOGIC CATACLYSMS
Revealing the Earth's Hazards, Revised Edition

For information contact:

Facts On File, Inc.
132 West 31st Street
New York NY 10001

Library of Congress Cataloging-in-Publication Data

Erickson, Jon, 1948–
Quakes, eruptions, and other geologic cataclysms : revealing the earth's hazards / Jon Erickson.—Rev. ed./foreword [by] Alexander E. Gates.
p. cm.—(The living earth)
Includes bibliographical references and index.
ISBN 0-8160-4516-X (acid-free paper)
1. Natural disasters. I. Title.

GB5014.E75 2001
550—dc21 2001023055

Facts On File books are available at special discounts when purchased in bulk quantities for businesses, associations, institutions or sales promotions. Please call our Special Sales Department in New York at 212/967-8800 or 800/322-8755.

You can find Facts On File on the World Wide Web at http://www.factsonfile.com

Text design by Cathy Rincon
Cover design by Nora Wertz
Illustrations by Jeremy Eagle and Dale Dyer, © Facts On File

Printed in the United States of America

8 7 6 5 4

on acid-free paper.

CONTENTS

TABLES

ACKNOWLEDGMENTS

The author thanks the National Aeronautics and Space Administration (NASA), the National Museums of Canada, the National Oceanographic and Atmospheric Administration (NOAA), the National Optical Astronomy Observatories (NOAO), the National Park Service, the U.S. Air Force, the U.S. Army, the U.S. Army Corps of Engineers, the U.S. Department of Agriculture (USDA), the USDA Forest Service, the USDA Soil Conservation Service, the U.S. Department of Energy, the U.S. Geological Survey (USGS), and the U.S. Navy for providing photographs for this book.

The author also thanks Frank K. Darmstadt, Senior Editor, and Cynthia Yazbek, Associate Editor, for their invaluable help with this project.

FOREWORD

Among the greatest fears of the human population in historical times has been the wrath of the Earth. Although science and engineering have effectively reduced both the surprise and devastating effects of such disasters, the possibility always exists of a geologic event that could overcome all of our careful preparation. With the burgeoning human population that spreads ever closer to areas of potential harm or packs areas of geologic instability, we could easily witness disasters of biblical proportions. Such disasters are the subject of countless books and movies. Many colleges and universities offer very well-attended courses on these disasters as well. This revised edition of *Quakes, Eruptions, and Other Geologic Cataclysms: Revealing the Earth's Hazards* by Jon Erickson explains and recounts these natural hazards in a single, understandable text. Surprisingly, very few such compilations exist.

The Earth can impose many distinct hazards on us. Some are cataclysmic whereas others are slow. Some can cause great loss of life whereas others cause great property damage but little loss of life. Some cause both. This book chooses to address nine distinct hazards that span the range of types. The first chapter explains the restless Earth and places several of the succeeding chapters into context. It explains the phenomenon of plate tectonics and those forces that serve to localize many of the hazards, particularly earthquakes and volcanoes. Chapter 2 recounts many of the intense and devastating earthquakes of our time. It describes the quakes, the damage inflicted, the associated phenomena, and the geologic setting of the faults that produced the quakes. The control on the location of seismically active zones and the recurrence of major earthquakes is placed into plate tectonic context. Chapter 3 recounts several catastrophic volcanic eruptions similar to the descriptions of earthquakes in chapter 2. It then explains the mechanics and chemistry of volcanic eruptions by including actual examples that illustrate each process. Finally, it places volcanoes into the plate tectonic model.

Chapters 4 through 10 describe phenomena that are gravitationally driven, related to climate change, or extraterrestrial. Chapters 4 and 5 on Earth movements and catastrophic collapse, respectively, are hazards that are gravity driven. They result from the failure of rock and soil that is precarious and unstable. These phenomena may be initiated by earthquakes, volcanoes, or neither. Many of the examples described are the result of earthquakes and volcanoes because they are the most spectacular. Chapters 6, 7, and 8 relate phenomena of climate. Floods are regular occurrences on virtually every river and in every coastal area. As the population encroaches flood-prone areas or as climate changes to increase precipitation in other areas, refining safeguards and methods of prediction becomes important. Desertification and associated dust storms result from overgrazing, certain farming techniques, and overuse of surface water and groundwater. All of these causes are common to modern civilization and increase with increasing population. Examples of the rapid expansion of the Sahara Desert southward and the Dust Bowl of the Midwest are used to exemplify this hazard. The final climatic hazard involves glaciation. Both the formation of continental glaciers and the melting thereof are processes of concern, especially with regard to coastal flooding.

The two final chapters include meteorite impacts and mass extinctions. The great dinosaur extinction is proposed to have resulted from a meteorite impact. However, several of the other topics may have also resulted in or aided in mass extinctions, namely volcanism and climate change. The human race is also responsible for a major mass extinction.

This book is one of surprisingly few compilations of several types of natural disasters. It is loaded with examples of specific events to explain the processes involved in their generation. This approach makes reading enjoyable to even the most fearful of science. It also contains a glossary to define the carefully chosen scientific terms that were used. This volume also contains a thorough bibliography for those who wish to read further. This book can be enjoyed by high school and college students alike besides those of us disaster enthusiasts.

—Alexander E. Gates, Ph.D.

INTRODUCTION

Geologic hazards have plagued people since time immemorial. We live on a dynamic planet, with devastating earthquakes, violent volcanoes, and other catastrophic geologic activities that destroy property and take human lives. These phenomena arise from the interactions of a jumble of crustal plates that make up the Earth's outer shell and that are constantly in motion. Tectonic forces are also responsible for raising mountains and creating a variety of geologic structures, often accompanied by earthquakes, volcanoes, and other earth-moving processes.

Earthquakes are the most destructive natural forces, producing widespread damage, destroying entire cities, and killing people by the thousands. Volcanoes are the next most destructive natural forces that destroy property and take many lives. Other geologic hazards include ground failures, floods, and dust storms. Floods are becoming much more hazardous because people crowd onto floodplains, which carry away excess water during river overflows. Dust storms can directly threaten life and cause severe soil erosion. The melting of the glaciers during a sustained warm climate could raise sea levels and drown coastal regions. As human populations continue to grow out of control on a planet with finite resources, we are placed into the perilous position of being one of the most destructive forces on Earth.

The text begins by examining the geologic forces that shape our planet. It then discusses the effects of ground shaking caused by earthquake faults. Next, volcanic activity and its dangers to civilization are investigated. The book continues with an examination of the geologic hazards posed by ground failures and catastrophic collapse. Flooding, perhaps the most pervasive of geologic hazards, is well presented. The effects of sand in motion in desert regions cannot be ignored as a major geologic hazard. The text also focuses on the effects of ice on the run, as glaciers melt and raise sea levels. Perhaps the most destructive geologic force is the impact of cosmic rubble onto the Earth. Our species causes another powerful environmental impact, as the world's plants and animals disappear at alarming rates.

This revised and updated edition is a much expanded examination of the geologic hazards faced by civilization. Science enthusiasts will particularly enjoy this fascinating subject and gain a better understanding of how the forces of nature operate on Earth. Students of geology and earth science will also find this a valuable reference to further their studies. Readers will enjoy this clear and easily readable text that is well illustrated with dramatic photographs, detailed illustrations, and helpful tables. A comprehensive glossary is provided to define difficult terms, and a bibliography lists references by chapter for further reading. The geologic processes that shape the surface of our planet are an example of the tireless forces that make this a living Earth.

1

THE DYNAMIC EARTH

PLATE TECTONIC ACTIVITY

This chapter examines the geologic forces that shape our world. The Earth is a highly dynamic planet, with rising mountain ranges, gaping canyons, erupting volcanoes, and shattering earthquakes. No other body in the solar system offers so many unusual landscapes, sculpted by highly active weathering agents that cut down tall mountains and gouge out deep ravines.

These activities are expressions of plate tectonics, from the Greek *tekton*, meaning "builder," which is the continual creation, motion, and destruction of sections of the Earth's crust. This action moves the continents around the globe, thus making the Earth a living planet both geologically and biologically. If not for a jumble of crustal plates interacting with each other to provide a myriad of geologic features, this would indeed be a desolate world.

THE NEW GEOLOGY

The Earth's outer shell is broken into seven major and about a half dozen minor crustal plates (Fig. 1) that are constantly in motion. The crustal plates

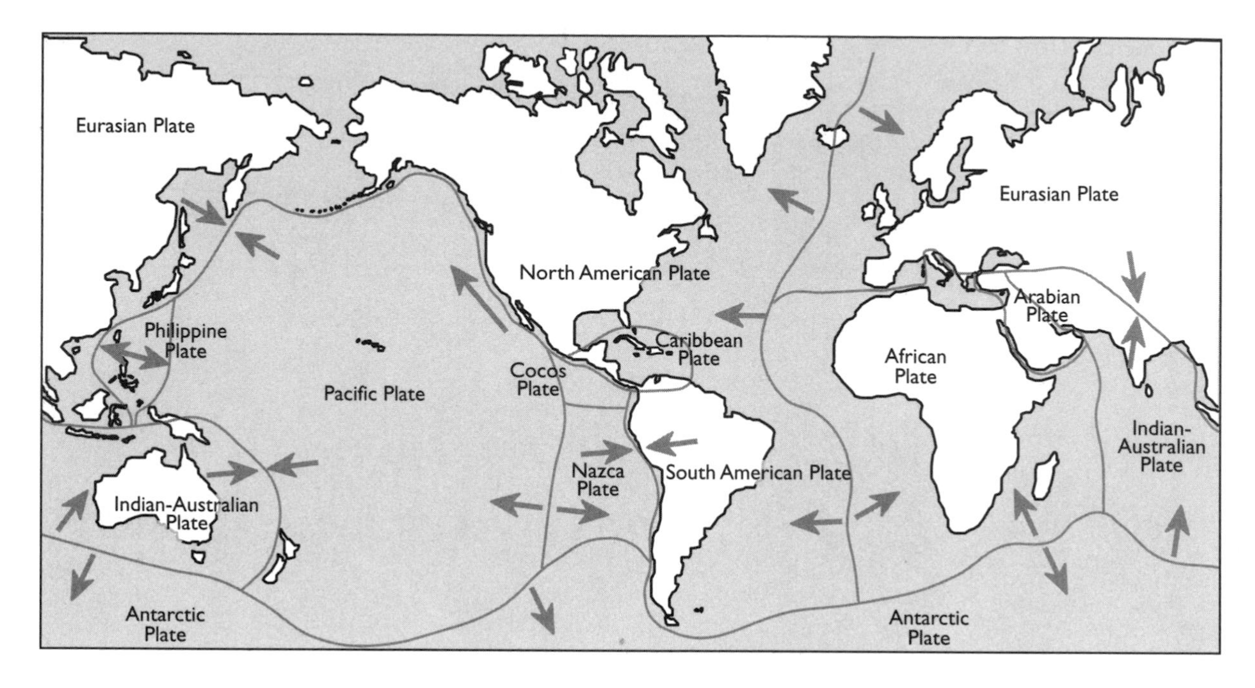

Figure 1 *The Earth's crust is fashioned out of several lithospheric plates that are responsible for the planet's active geology.*

ride on a hot, pliable outer layer of the mantle, called the asthenosphere. Interactions among the various plates shape the surface of the planet. The shifting plates range in size from a few hundred to tens of millions of square miles and average about 60 miles thick. This structure is unique and important for the operation of plate tectonics, which is largely responsible for the Earth's active geology.

The lithospheric plates meet at two different kinds of intersections: divergent plate margins, where plates pull away from each other, and convergent plate margins, where they collide with one another. Narrow midocean ridges and oceanic trenches sharply mark the edges of crustal plates. However, ocean plate boundaries are much broader, up to thousands of miles wide. Where two plates move toward each other, the crust folds up. Where two plates move away from each other, the crust stretches and thins out. An analogy would be a loosely fitting jigsaw puzzle with large gaps between pieces.

The divergent plate margins comprise long chains of volcanoes on the deep ocean floor. There basalt welling up from within the mantle creates new oceanic crust in a process known as seafloor spreading. More than 4.3 cubic miles of new rock are produced by this process each year. The midocean ridge system snakes around the globe like the stitching on a baseball for a distance of 40,000 miles, thus making it the longest uninterrupted structure on Earth (Fig. 2). Some molten magma erupts onto the surface of the ridge as lava, but the vast majority cools and bonds to the edges of separating plates. Periodically, molten rock overflows onto the ocean floor in gigantic eruptions, providing several square miles of new oceanic crust yearly.

The oceanic crust and underlying lithosphere sink into the mantle at convergent plate margins called subduction zones, which create deep-sea trenches in the ocean floor. When two plates collide, the less buoyant oceanic crust subducts under continental crust or younger oceanic crust, which due to its higher temperature is less dense. The line of subduction is marked by the deepest trenches in the world.

As old oceanic crust plunges into the mantle, it melts to provide new basalt in a continuous cycle of crustal generation. The magma produced in this manner eventually reaches the surface. There it causes spectacular explosive eruptions along chains of volcanoes adjacent to the subduction zones called island arcs. If tied end to end, the subduction zones would stretch completely around the world. The subduction of seafloor into the underlying mantle along the western and northern borders of the Pacific typically occurs at a rate of 4 inches per year. However, at the Tonga Trench, in the South Pacific just east of Fiji, the crustal plate carrying Australia is diving into the 35,000-foot-deep chasm at a record rate of more than 9 inches a year.

The deep trenches created by descending plates accumulate large deposits of sediments derived from the adjacent continents and island arcs. When the plate dives into the interior, most of its waterlogged sediments go down with it. Yet much more water is being subducted into the Earth than is coming out of subduction-zone volcanoes. Heat and pressure act to dehydrate

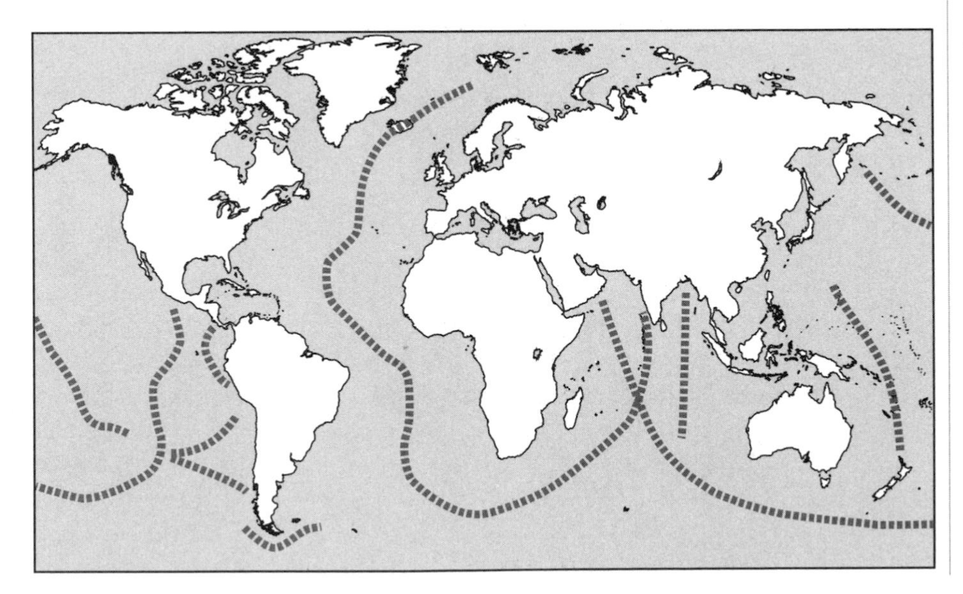

Figure 2 *Midocean ridges, where crustal plates are spreading apart, comprise the most extensive mountain chains in the world and are centers of intense volcanic activity.*

rocks of the descending plate. However, just where all the fluid goes has remained a mystery. Some fluid expelled from a subducting plate reacts with mantle rocks to produce low-density minerals that slowly rise to the seafloor. There they build mud volcanoes that erupt serpentine, an asbestos mineral formed by the reaction of olivine from the mantle with water.

Oceanic crust created at spreading ridges is destroyed in subduction zones, which is responsible for moving the lithospheric plates around the surface of the Earth (Fig. 3). Most subduction zones surround the Pacific Basin. Plate subduction generates the intense seismic activity that fringes the Pacific Ocean in a region known as the circum-Pacific belt. It is synonymous with the Ring of Fire, known for its extensive volcanism. As the oceanic crust descends into the mantle, it remelts to provide new molten magma for volcanoes bordering the subduction zones. These volcanoes form long chains of islands, mostly in the Pacific, and most volcanic mountains on the continents.

The lithospheric plates carry continental crust around the surface of the Earth like ships frozen in Arctic pack ice. The bulk of the crust comprises granitic and metamorphic rocks, which constitute most of the continents. Because the continental crust contains light materials, it remains on the surface due to its greater buoyancy. Nevertheless, plate tectonics has recycled significant amounts of continental crust over the last 4 billion years as subduction carries pieces of continents into the mantle. During this time, the floors of as many as 20 oceans have disappeared into the mantle as well.

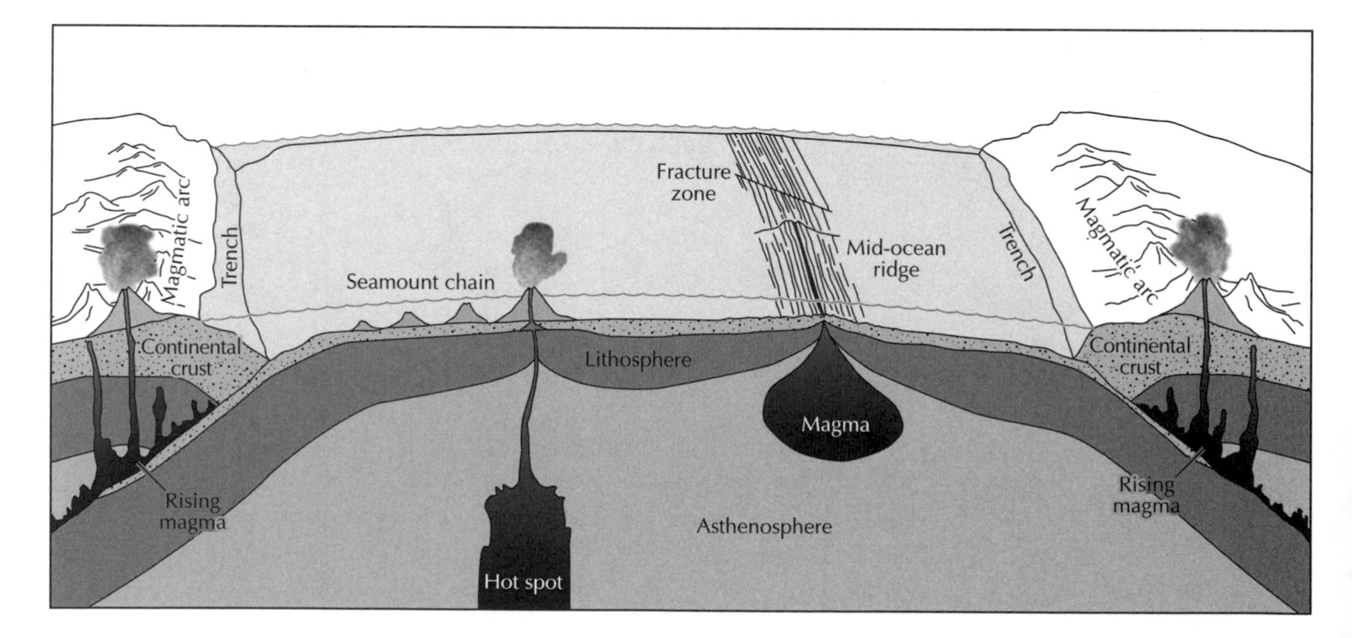

Figure 3 *Midocean ridges and subduction zones are responsible for moving lithospheric plates around the surface of the Earth.*

Figure 4 *The explosive eruption of Mount St. Helens, Washington, on May 18, 1980.*
(Courtesy USDA)

Plate collisions uplift mountain ranges on the continents and create volcanic islands on the ocean floor. When an oceanic plate subducts beneath a continental plate, it forms sinuous mountain chains, such as the Andes of South America, and volcanic mountain ranges, such as the Cascades of the American Pacific Northwest. This range is home to Mount St. Helens, which produced one of the most powerful eruptions on the North American continent (Fig. 4). The breakup of a plate creates new continents and oceans, and the collision of plates builds supercontinents. The process of rifting and patching of the continents has been an ongoing process for most of the Earth's existence.

The continents most recently rifted apart and drifted away from each other about 180 million years ago (Fig. 5), when a supercontinent called Pangaea, from Greek meaning "all lands," rifted apart along the present Mid-

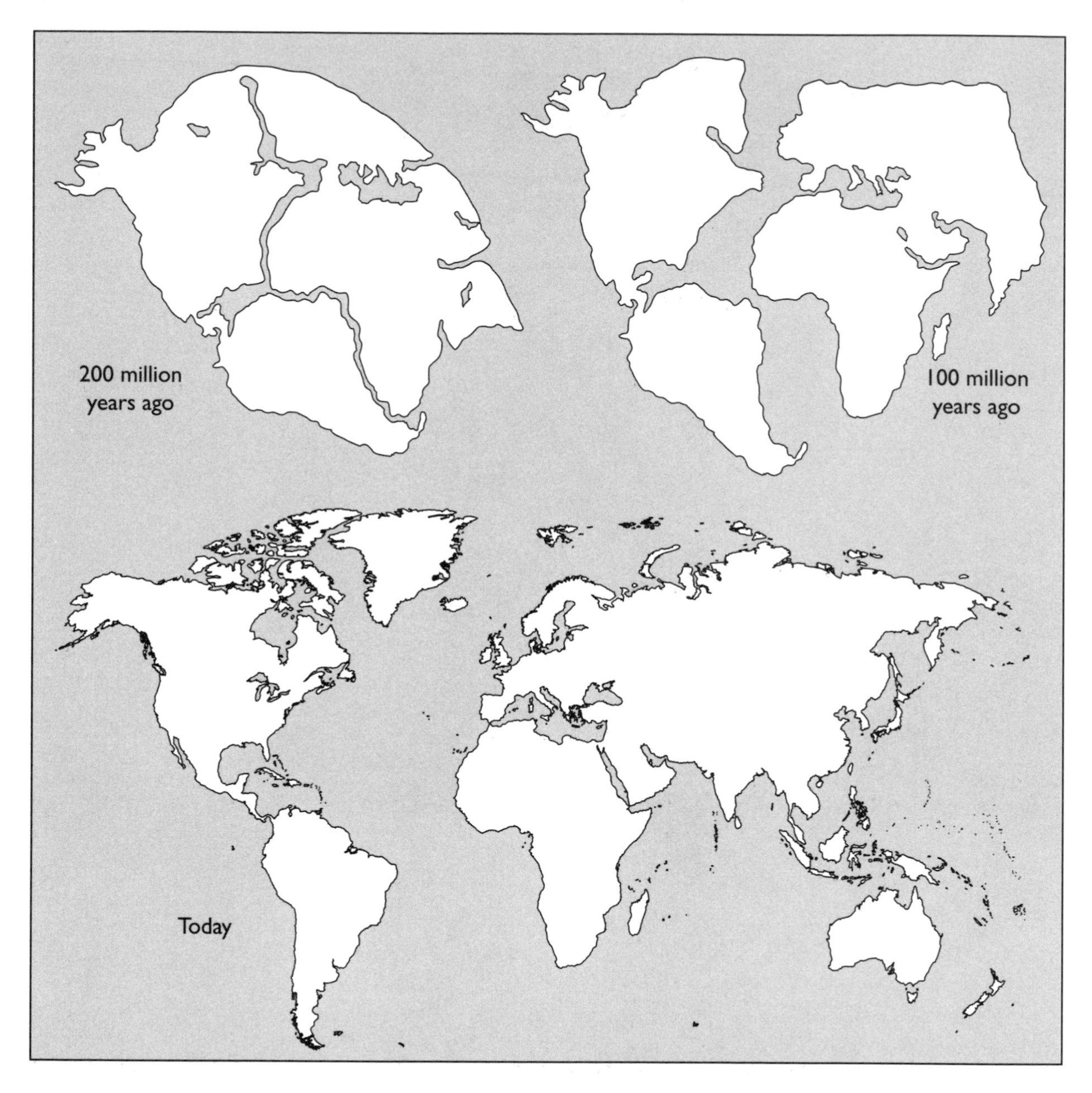

Figure 5 *The breakup continents. About 180 million years ago, a supercontinent called Pangaea rifted apart to form the present-day continents.*

Atlantic Ridge. Upwelling magma created an undersea volcanic mountain range that runs through the middle of the Atlantic Ocean, weaving halfway between the continents surrounding the Atlantic Basin. The Mid-Atlantic Ridge is part of a global spreading ridge system responsible for creating new oceanic crust. Molten magma rising from deep within the mantle forms new lithosphere by the addition of basalt to plate margins.

The two lithospheric plates of the Atlantic Basin spread apart an inch or more per year. As the Atlantic Basin widens, the surrounding continents separate, thereby compressing the Pacific Basin. The Pacific Basin is ringed by subduction zones that assimilate old lithosphere, causing the Pacific plate along with adjacent plates to shrink. The Pacific plate, the world's largest lithospheric plate at nearly 7,500 miles across, originated as a microplate no bigger than the United States some 190 million years ago. It achieved its present size by the gradual addition of lithosphere at associated spreading ridges.

The oceanic crust, composed of basalts originating at spreading ridges and sediments washed off nearby continents, is uniformly about 3 to 5 miles thick. As the newly formed plate moves away from the ridge and cools, the lithosphere thickens with age. After about 60 million years, it reaches a thickness of about 60 miles, increases in density, and eventually subducts into the mantle.

On its journey deep into the Earth's interior, the lithosphere and its overlying sediments melt. The molten magma rises toward the surface in huge bubblelike structures called diapirs, from the Greek *diapeirein,* meaning "to pierce." When the magma reaches the base of the crust, it provides new molten rock for magma chambers beneath volcanoes and granitic bodies called plutons, which often form mountains. In this manner, plate tectonics continuously changes and rearranges the face of the Earth.

THE ACTIVE CRUST

The crust comprises less than 1 percent of the Earth's radius and about 0.5 percent of its mass. It is composed of ancient continental rocks and comparatively young oceanic rocks. As much as 70 percent of the continental crust emerged during the Archean eon, with a major episode of growth between 3.0 and 2.5 billion years ago. The continental crust resembles a layer cake with sedimentary rocks on top, granitic and metamorphic rocks in the middle, and basaltic rocks on the bottom. This structure resembles a jelly sandwich, with a pliable middle layer placed between a solid upper crust and a hard lithosphere, the rigid uppermost layer of the mantle. Most of the continental rock originated from volcanoes stretching across the ocean drawn together by plate tectonics.

Underlying all other rocks on the Earth's surface is a thick layer of basement complex, composed of ancient granitic and metamorphic rocks that have been in existence for 90 percent of Earth history. These rocks form the nuclei of the continents. They first appeared during a period of mantle segregation and outgassing, which created the crust along with the atmosphere and ocean. One remarkable feature about these rocks is that despite their great age, they are similar to more recent rocks. This signifies that geologic processes began quite early in geologic history and have had a long and productive life.

The continental crust covers about 45 percent of the Earth's surface, when including continental margins and small, shallow regions in the ocean. It varies from 6 to 45 miles thick and rises on average about 2.7 miles above the ocean floor or about 4,000 feet above sea level. The thinnest parts of the continental crust lie below sea level on continental margins. The thickest portions underlie towering mountain ranges, which are underlain by deep crusts called crustal roots. For example, the highest ranges in the world, whose tallest peaks are more than 4 miles high, are supported by crustal roots 40 to 50 miles thick. Below the continental crust, subcrustal keels composed of relatively cool attached mantle material as much as 250 miles thick ride along with the continents.

Basement rocks deep below the surface form the nuclei around which the continents grew. They are exposed in broad, domelike structures called shields (Fig. 6), which contain the oldest rocks on Earth. They are extensive uplifted areas essentially bare of recent sedimentary deposits and contain only thin soils. Surrounding the shields are broad, shallow depressions of basement rock covered with nearly flat-lying sedimentary strata called continental platforms.

The best-known of the dozen or so continental shields are the Canadian Shield in North America and the Fennoscandian Shield in Europe. The exposure of the Canadian Shield from Manitoba to Ontario is attributed to uplifting of the crust by a mantle plume and the erosion of sediments in the uplifted area. Some of the oldest-known rocks of North America are the 2.5-billion-year-old

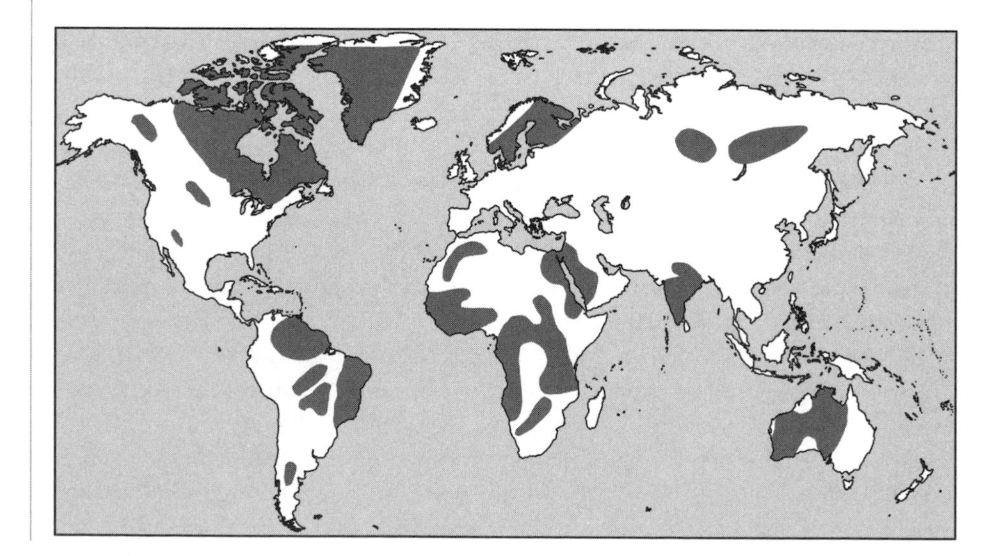

Figure 6 *Areas of exposed Precambrian shields in the continental interiors.*

granites of the Canadian Shield. One of the most ancient rocks is the Acasta Gneiss in northwest Canada, which originally formed deep within the crust some 4 billion years ago and now outcrops to the surface. These regions are fully exposed where flowing ice sheets eroded their sedimentary cover during the Pleistocene ice ages. More than a third of Australia is Precambrian shield, and sizable shields lie in the interiors of Africa, South America, and Asia as well.

The shields are interspersed with greenstone belts comprising a mixture of metamorphosed lava flows and sediments derived from volcanic island arcs caught between colliding continents. Indeed, most of the continental rock present on Earth has been produced by the paving action of volcanic arcs operating for billions of years. The very ancient centers of continents are made up, in large part, of Archean island arcs. Greenstones have no modern equivalents. Therefore, the geologic conditions under which they formed were very different from those observed today.

Greenstone belts span an area of several hundred square miles and are surrounded by immense expanses of gneiss, the metamorphic equivalent of granite and the predominant Archean rock type formed between 4 and 2.5 billion years ago. Their color derives from chlorite, a greenish, micalike mineral. The best-known greenstone belt is the Swaziland sequence in the Barberton Mountain Land of southeastern Africa. It is nearly 12 miles thick and more than 3 billion years old.

Geologists are especially interested in greenstone belts because they contain most of the world's gold. India's Kolar greenstone belt holds the richest gold deposits. It is some 3 miles wide and 50 miles long. It formed when two plates clashed about 2.5 billion years ago. In Africa, the best deposits are in rocks as old as 3.4 billion years, and most South African gold mines are found in greenstone belts. In North America, the best gold mines are in the Great Slave Province of northwest Canada, where well over 1,000 deposits are known. The Great Slave is an extremely ancient region of North America that formed sometime between 4.0 and 2.5 billion years ago.

Scattered throughout the greenstone belts are ophiolites, from the Greek word *ophis,* meaning "serpent." They are slices of ocean floor shoved onto the continents by drifting plates and range in age up to 3.6 billion years old. Pillow lavas (Fig. 7), tubular bodies of basalt erupted on the ocean floor, also appear in the greenstone belts. These deposits are among the best evidence for plate tectonics operating early in the Precambrian. Therefore, continents have been on the go practically from the very beginning. Many ophiolites also contain rich ores that provide important mineral resources the world over. These include the Apennines of northern Italy, the Urals of Russia, and the Andes of South America.

Blueschists (Fig. 8) were also thrust onto the continents, providing further evidence for early tectonic activity. They are metamorphosed rocks of

Figure 7 *Pillow lava exposed on the south side of Wadi Jizi northeast of Suhaylah, Sultanate of Oman.*

(Photo by E. H. Bailey, courtesy USGS)

subducted oceanic crust forced into the mantle at subduction zones on the ocean floor and formed under high pressure. The continental collisions resulted in linear formations of greenish volcanic rocks along with light-colored masses of granite and gneiss, common igneous and metamorphic rocks making up the bulk of the continents.

Figure 8 *An outcrop of retrograde blueschist rocks in Nome District, Seward Peninsula, Alaska, represents subducted oceanic crust.*

(Photo by C. L. Sainsbury, courtesy USGS)

The shields and their surrounding basement rock, called platforms, together comprise the stable cratons. These lie in the continental interiors and were the first pieces of land to appear. Cratons consist of ancient igneous and metamorphic rocks, whose composition is remarkably similar to their modern equivalents. The existence of cratons early in Earth history suggests a fully operating rock cycle was already in place by this time.

Cratons are a patchwork of crustal blocks combined into geologic collages known as terranes. They are usually bounded by faults and are distinct from their geologic surroundings. The boundaries between two or more terranes are called suture zones, consisting of ancient oceanic crust shoved onto the continents by drifting plates. The composition of most terranes is similar to that of a volcanic island or undersea plateau. Others comprise a consolidated conglomerate of pebbles, sand, and silt that accumulated in an ocean basin between colliding crustal fragments.

Terranes are generally elongated bodies that deformed when colliding and accreting to a continent. The assemblage of terranes in China is being stretched and displaced in an east-west direction as India, itself a single great terrane, continues to press against southern Asia after colliding with the mainland some 45 million years ago. The buckling crust raised the Himalaya Mountains and the broad Tibetan Plateau, the largest topographic upland on Earth. In the process, Asia was squeezed to accommodate the northward advancement of India.

For tens of millions of years, India rode northward across the equatorial Tethys Sea separating Africa and Eurasia on its underlying tectonic plate at a rate of about 4 inches per year. As the Tethys seafloor met Asia, it slid beneath the continent, subducted into the Earth's mantle, and melted to provide magma for volcanoes. Like a bulldozer, the edge of Asia scraped sediment off the seafloor and piled it into a great heap. By about 35 million years ago, India butted head-on into the sediment heap and began to raise the Himalayas by folding and by faulting. This stacked huge slices of rock onto one another. The faults, as they still do, generated massive earthquakes.

Granulite terranes are high-temperature metamorphic belts, which formed in the deeper parts of continental rifts. They also comprise the roots of mountain belts formed by continental collision, such as the Alps and Himalayas. North of the Himalayas is a belt of ophiolites, which marks the boundary between the sutured continents. Terrane boundaries are commonly marked by ophiolite belts, consisting of marine sedimentary rocks, pillow basalts, sheeted dike complexes, gabbros, and peridotites.

Terranes come in a variety of shapes and sizes ranging from small slices of crust to subcontinents as large as India. They range in age from well over a billion to less than 200 million years old. The ages of the terranes were determined by studying entrained fossil radiolarians (Fig. 9). These are marine pro-

Figure 9 *Upper Devonian radiolarians separated from chert of the Ford Lake Shale, Kandik Basin, Eagle District, Yukon region, Alaska.*

(Photo by D. L. Jones, courtesy USGS)

tozoans with skeletons made of silica abundant from about 500 million to 160 million years ago. Different species also define specific regions of the ocean where the terranes originated.

Most terranes comprise fault-bounded blocks whose geologic histories are different from those of neighboring terranes and of adjoining continental masses. Many have traveled considerable distances before finally colliding with continental margins. Some North American terranes originated in the western Pacific and traversed thousands of miles eastward. Large chunks of land also traveled from Mexico to Canada and Alaska. For instance, some 70 million years ago, Vancouver Island in the Pacific Northwest was once nestled along the coast of what is now Baja California. In a similar fashion, Eurasia, the largest continent, is accumulating pieces of crust arriving from the south.

The accreted terranes played a major role in the formation of mountain chains along convergent continental margins that mark the collision of lithospheric plates. Geologic activity around the Pacific Rim was responsible for practically all mountain ranges facing the Pacific Ocean and the island arcs along its perimeter. Along the mountain ranges in western North America, the terranes are elongated bodies due to the slicing of the crust by a network of northwest-trending faults. The most active of these faults is California's San Andreas, which has been displaced some 200 miles over the last 25 million years (Fig. 10).

The Gulf of California, separating the Baja California Peninsula from mainland Mexico, is a continuation of the San Andreas Fault system. The landscape is literally being torn apart while opening one of the youngest and richest seas on Earth. It began rifting some 6 million years ago, offering a new outlet to the sea for the Colorado River. Subsequently, the Colorado River began carving out the mile-deep Grand Canyon.

MANTLE CONVECTION

All geologic activity taking place on the surface of the planet is an outward expression of the great heat engine in the Earth's interior. Although the mantle is solid rock, the intense heat causes it to flow slowly. The circulation of heat within the mantle is the main driving force behind plate tectonics. Convection currents and mantle plumes (columns of molten magma) transport heat from the core to the underside of the lithosphere. This process is responsible for volcanic activity on the ocean floor and on the continents. Most mantle plumes originate from within the mantle. However, some arise from the very bottom of the mantle, like huge bubbles rising from deep within the Earth.

Figure 10 *A stream in the Carrizo Plains has been offset 0.25 miles across the San Andreas Fault as the land to the west moves northward.*

(Photo by R. E. Wallace, courtesy USGS)

Convection is the motion within a fluid medium resulting from a difference in temperature from top to bottom. Fluid rocks in the mantle acquire heat from the core, ascend, dissipate heat to the lithosphere, cool, and descend back to the core for more heat. Lithospheric plates, created at spreading ridges and destroyed at subduction zones, are the products of convection currents in the mantle.

The formation of molten rock in the mantle and the rise of magma to the surface is due to an exchange of heat within the planet's interior. The Earth is steadily losing heat from the mantle to the surface through its outer shell, or lithosphere. Some 70 percent of this heat loss results from seafloor spreading, and most of the rest is due to volcanism at subduction zones. However, volcanic eruptions represent only localized and highly spectacular releases of this heat energy (Fig. 11).

The mantle has cooled to a semisolid or plastic state, except for a relatively thin layer of partially melted rock between 70 and 150 miles below the surface, called the asthenosphere. The asthenosphere mechanically acts as a fluid and flows by convection. Heat transferred from the mantle to the asthenosphere produces convective currents that rise and travel laterally upon reaching the underside of the lithosphere. After giving up heat to the lithosphere, the currents cool and descend back into the mantle.

Most of the Earth's thermal energy is generated by radioactive isotopes, mainly potassium (K), uranium (U), and thorium (Th), abbreviated KUT. The temperature within the Earth increases rapidly with depth. At a depth of about 70 miles, where the material of the upper mantle begins to melt, the temperature is about 1,200 degrees Celsius. This marks the semimolten region of the upper mantle, or the asthenosphere, on which the rigid lithospheric plates ride. A continent essentially floats on the denser, partially molten asthenosphere, with much of its bulk below the surface like an iceberg on the high seas.

As continents gain mass through lava flows, granitic intrusions, and sedimentation, they press down deeper into the asthenosphere. The asthenosphere is constantly losing material, which adheres to the undersides of lithospheric

Figure 11 *Molten lava pouring into the sea from an eruption of Kilauea Volcano, Hawaii.*

(Photo by R. T. Haugen, courtesy National Park Service)

plates. If the asthenosphere were not continuously fed new material from mantle plumes, the plates would grind to a halt. The Earth would then become, in all respects, a dead planet. Fortunately, this event is not expected for several billion years.

The temperature of the upper mantle increases gradually to about 2,000 degrees Celsius at a depth of 300 miles and then rapidly increases to the top of the core, where temperatures reach 5,000 degrees. The lower mantle begins at a depth of about 400 miles and is composed of primitive rock that has not changed significantly since the Earth's infancy. By comparison, the upper mantle has lost much of its lighter components and volatiles, which have congregated in the crust, ocean, and atmosphere.

Most of the mantle's heat is generated internally by radiogenic sources. The rest is supplied by the core, which has retained much of its heat since the early accretion of the Earth some 4.6 billion years ago. The temperature difference between the mantle and the core approaches 1,000 degrees. Material from the mantle might mix with the fluid outer core to form a distinct layer on its surface that could block heat flowing from the core to the mantle and interfere with mantle convection.

The surface of the Earth is continuously being shaped by the action of the mantle churning below the lithosphere (Fig. 12). The mantle currents travel very

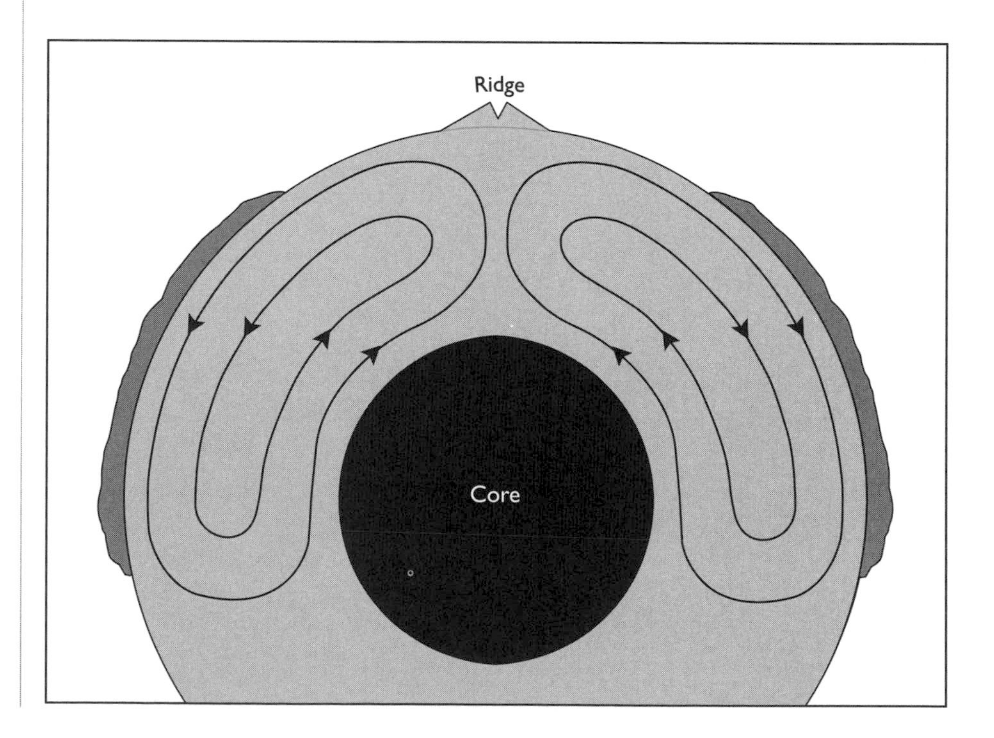

Figure 12 *Convection currents in the mantle move the continents around the Earth.*

slowly, completing a single convection loop in perhaps hundreds of millions of years. Without mantle convection, erosion would wear down mountains to the level of the prevailing plain in a matter of only 100 million years, or merely 2 percent of the Earth's age. The surface of the Earth would then become a vast, featureless plain, unbroken by mountains and valleys. No volcanoes would erupt, and no earthquakes would rumble across the land. This would leave the planet as geologically and biologically dead as the Moon.

The mantle convection cells act like rollers beneath a conveyor belt to propel the lithospheric plates forward. Hot material rises from within the mantle and circulates horizontally near the Earth's surface. The top layer cools to form the rigid lithospheric plates, which carry the crust around the surface of the planet. The plates complete the mantle convection by plunging back into the Earth's interior. In this manner, they are simply surface expressions of mantle convection. If fractures or zones of weakness appear in the lithosphere, the convection currents spread the fissures wider apart to form rift systems. This is where the Earth loses the largest portion of its interior heat to the surface, as magma flows out of the rift zones to form new oceanic crust.

SEAFLOOR SPREADING

Seafloor spreading is described as the wound that never heals as magma continuously rises to the Earth's surface. This creates new lithosphere at spreading ridges on the ocean floor and generates more than half of the Earth's crust. Seafloor spreading begins with hot rocks rising by convection currents in the mantle. After reaching the underside of the lithosphere, the mantle rocks spread out laterally, cool, and descend back into the Earth's interior. The constant pressure against the bottom of the lithosphere fractures the plate and weakens it.

As the convection currents flow outward on either side of the fracture, they carry the separated parts of the lithosphere along with them, widening the gap. The rifting reduces the pressure, allowing mantle rocks to melt and rise through the fracture zone. The molten rock passes through the lithosphere and forms magma chambers that supply molten rock for the generation of new lithosphere. The greater the supply of magma to the chambers, the higher the overlying spreading ridge is elevated.

The magma flows outward from a trough between ridge crests and adds new layers of basalt to both sides of the spreading ridge, thereby creating new lithosphere. The continents are carried passively on the lithospheric plates created at spreading ridges and destroyed at subduction zones. Therefore, the engine that drives the birth and evolution of rifts and, consequently, the breakup of continents and the formation of oceans ultimately originates in the mantle.

The new lithosphere created at spreading ridges starts out thin and eventually thickens by the underplating of magma from the upper mantle along with the accumulation of overlying sediment layers. As the lithosphere moves away from the spreading ridge, it cools and becomes thicker and denser. The segment near continental margins, where the ocean is the deepest, is about 60 miles thick. Eventually, the lithosphere becomes so heavy it can no longer remain on the surface and sinks into the mantle, forming a deep trench at the point of subduction.

The mantle rocks beneath spreading ridges that create new lithosphere consist mostly of peridotite, an iron-magnesium silicate. As the peridotite melts on its journey through the lithosphere, a portion becomes highly fluid basalt, the most common magma erupted onto the Earth's surface. About 5 cubic miles of new basalt adds to the crust yearly, mostly on the ocean floor at spreading ridges. The rest contributes to the continued growth of the continents.

The spreading ridges are centers of intense seismic and volcanic activity. This activity manifests itself as high heat flow from the Earth's interior. The greater the flow of magma to the ridge crest, the more rapid the seafloor spreading. The spreading ridges in the Pacific Ocean are more active than those in the Atlantic and therefore have less relief. Rapid-spreading ridges do not achieve the heights of slower ones because magma has less opportunity to pile up into tall heaps like those in the Atlantic.

The spreading ridge system does not form a continuous line. Instead, it is broken into small, straight sections called spreading centers (Fig. 13). The

Figure 13 *Spreading centers on the ocean floor are separated by transform faults.*

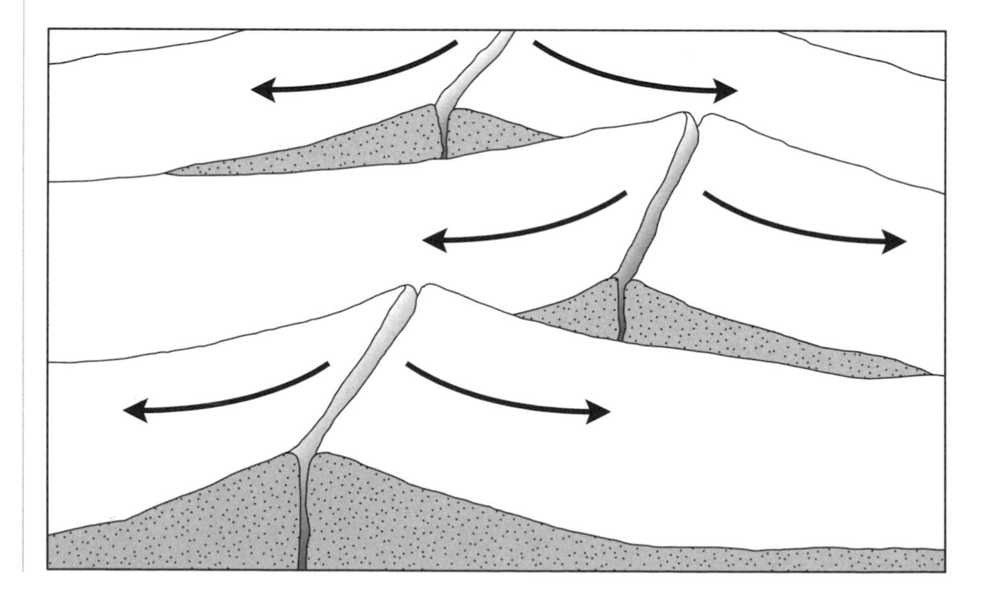

movement of new lithosphere generated at the spreading centers produces a series of fracture zones. These are long, narrow, linear regions up to 40 miles wide that consist of irregular ridges and valleys aligned in a stair step shape. When lithospheric plates slide past each other as the seafloor spreads apart, they create transform faults that range in size from a few miles to several hundred miles long. The faults occur every 20 to 60 miles along the Mid-Atlantic Ridge. Transform faults result from lateral strain, which is how movable lithospheric plates should react on the surface of a sphere. This activity appears to be more intense in the Atlantic, where the spreading ridge system is steeper and more jagged than in the Pacific and Indian Oceans.

Transform faults of the Mid-Atlantic Ridge generally have more relief than those of the 6,000-mile-long East Pacific Rise. This is the great spreading ridge system of the Pacific and the counterpart to the Atlantic ridge system. Fewer widely spaced transform faults exist along the East Pacific Rise, which marks the boundary between the Pacific and Cocos plates. Furthermore, the rate of seafloor spreading is 5 to 10 times faster for the East Pacific Rise than for the Mid-Atlantic Ridge.

The East Pacific Rise hosts exotic chimneys called black smokers that spew hot water blackened with sulfide minerals. The hot water originates from deep below the surface. There seawater percolating down through cracks in the oceanic crust comes into contact with magma chambers below the spreading centers. The hot water rising to the surface expels through hydrothermal vents. Living among the vents are clusters of tall tube worms, giant crabs, huge clams up to a foot long, and clusters of mussels, creatures the likes of which exist nowhere else on Earth.

SUBDUCTION ZONES

The spreading seafloor in the Atlantic is offset by the shrinking of the ocean floor in the Pacific. Deep trenches ring the Pacific Basin (Fig. 14), where old lithosphere subducts into the mantle. The subduction zones lying off continental margins and island arcs are regions of intense volcanic activity that produces some of the most explosive volcanoes on Earth. The volcanic island arcs that fringe the subduction zones share similar curved patterns. This curvature is the geometric shape that develops when a plane cuts a sphere, for example when a rigid lithospheric plate subducts into the mantle.

The subduction zones are also sites of almost continuous seismic activity deep in the bowels of the Earth. A band of earthquakes marks the boundaries of a sinking lithospheric plate. As plates press against each other along dipping fault planes, they create compressional earthquakes that can be highly destructive. Such earthquakes have continuously plagued Japan and the Philip-

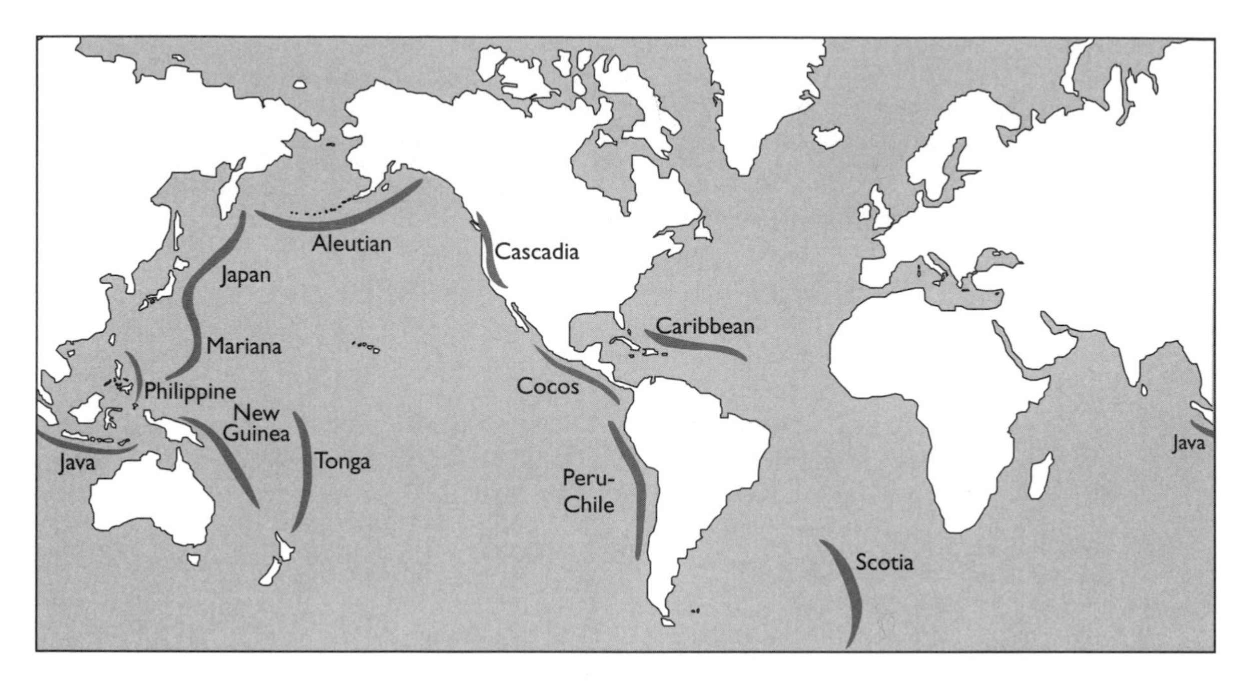

Figure 14 *Subduction zones, where crustal plates are forced into the Earth's interior, are marked by deep-sea trenches.*

pines as well as other islands connected with subduction zones. Many thousands of islands lying between Australia and Eurasia that make up the Indonesian and Philippine archipelagos will eventually accrete to the edge of Asia as huge tectonic plates converge.

Lithospheric subduction plays a fundamental role in plate tectonics and accounts for many of the geologic processes that shape the planet. As a plate cools, it grows thicker and denser by a process known as underplating, in which magma from the asthenosphere adheres to the underside of the plate. When the plate becomes thick and heavy, it loses buoyancy and sinks into the mantle at clearly defined subduction zones. As the plate descends, it drags the rest of the plate along with it somewhat like a locomotive pulling a freight train. Therefore, plate subduction is the main force behind plate tectonics. Pull at subduction zones is favored over push at spreading ridges to drive the continents around the surface of the globe.

Plate subduction in the Pacific forms some of the deepest trenches in the world (Table 1). As a lithospheric plate extends away from its place of origin at a spreading ridge, it cools and thickens as more material from the asthenosphere sticks to its underside. Eventually, the plate becomes so dense it loses buoyancy and sags deeper into the mantle. The depth at which a lithospheric plate sinks as it moves away from a spreading ridge varies with the plate's age. The older the lithosphere, the more basalt has adhered and the deeper the plate. Crust that is 2 million years old lies about 2 miles deep; crust that is 20

TABLE 1 THE WORLD'S OCEAN TRENCHES

Trench	Depth (miles)	Width (miles)	Length (miles)
Peru-Chile	5.0	62	3,700
Java	4.7	50	2,800
Aleutian	4.8	31	2,300
Middle America	4.2	25	1,700
Mariana	6.8	43	1,600
Kuril-Kamchatka	6.5	74	1,400
Puerto Rico	5.2	74	960
South Sandwich	5.2	56	900
Philippine	6.5	37	870
Tonga	6.7	34	870
Japan	5.2	62	500

million years old lies about 2.5 miles deep; and crust that is 50 million years old lies about 3 miles deep.

As the lithospheric plate sinks into the mantle, the line of subduction creates a deep trench that accumulates large amounts of sediment derived from an adjacent continent or island arc. The continental shelf and slope contain thick deposits of sediment washed off the nearby continent. When the sediments and their seawater content are caught between a subducting oceanic plate and an overriding continental plate, they are subjected to strong deformation, shearing, heating, and metamorphism (recrystallization without melting). The sediments are carried deep into the mantle, where they melt to become the source of new magma for volcanoes that fringe the subduction zones.

When the magma reaches the surface, it erupts onto the ocean floor, creating new volcanic islands. Most volcanoes, however, do not rise above sea level and instead became isolated undersea volcanoes called seamounts. The Pacific Basin is more volcanically active and has a higher density of seamounts than the Atlantic or Indian Basins. The number of undersea volcanoes increases with increasing crustal age and thickness. The tallest seamounts rise more than 2.5 miles above the seafloor and exist in the western Pacific near the Philippine Trench, where the oceanic crust is more than 100 million years old. The average density of Pacific seamounts is as much as 10 volcanoes per 5,000 square miles of seafloor, a considerably larger number of volcanoes than exists on the continents.

Figure 15 *The September 23, 1952, submarine eruption of Myojin-sho Volcano, Izu Islands, Japan.*
(Courtesy USGS)

Subduction-zone volcanoes are highly explosive (Fig. 15) because their magmas contain a large quantity of volatiles and gases that escape violently when reaching the surface. The type of volcanic rock erupted in this manner is called andesite. It is named for the Andes Mountains that form the spine of South America and are well known for their explosive eruptions.

The seaward boundaries of the subduction zones are marked by deep trenches. These lie at the edges of continents or along volcanic island arcs. Subduction zones, where cool, dense lithospheric plates dive into the mantle, are regions of low heat flow and high gravity. Conversely, the associated island arcs are regions of high heat flow and low gravity due to volcanism.

Behind the island arcs are marginal or back-arc basins that form depressions in the ocean floor due to plate subduction. Like island arcs, back-arc basins are also regions of high heat flow because of the upwelling of magma from deep-seated sources. Deep subduction zones, such as the Mariana Trench in the western Pacific, form back-arc basins. The Mariana Trench, which reaches a depth of nearly 7 miles, is the world's deepest and forms a long line northward from the island of Guam. Shallow subduction zones, such as the Chilean Trench off the west coast of South America, do not form back-arc basins. The back-arc basin that forms the Sea of Japan between China and the Japanese archipelago, which is a combination of ruptured continental fragments, will eventually be squeezed dry as the islands are plastered against Asia due to the interactions of lithospheric plates.

PLATE INTERACTIONS

The continental crust is broken up by faults that often leave long, linear structures when exposed on the surface. The continental crust and underlying lithosphere are generally between 50 and 100 miles thick. Therefore, breaking open a continent would appear to be a monumental task. During the rifting of continents into separate plates (Fig. 16), thick lithosphere must somehow give way to thin lithosphere. The transition from a continental rift to an oceanic rift is accompanied by block faulting. Huge blocks of continental crust drop down along extensional faults, where the crust is diverging. Convection currents rising through the mantle spread out under either side of the

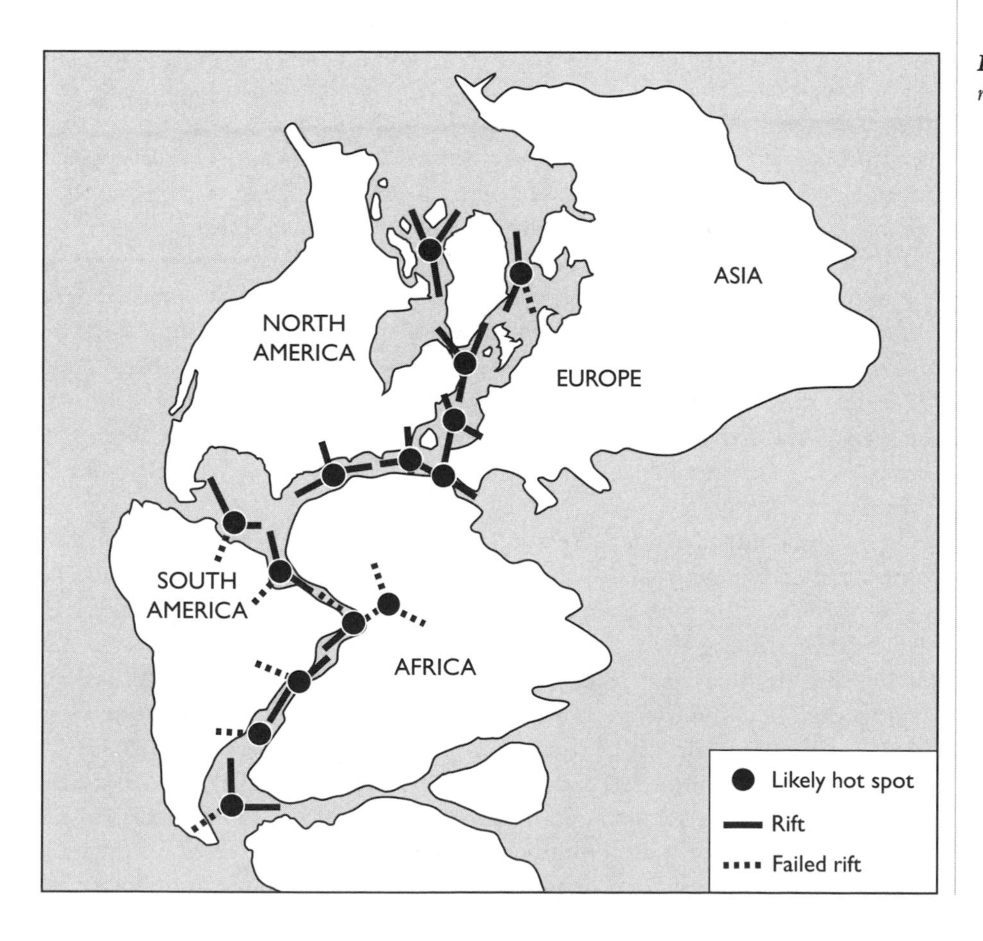

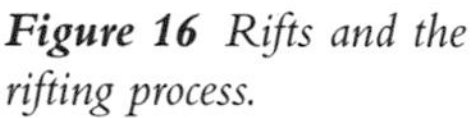
Figure 16 *Rifts and the rifting process.*

lithosphere, pulling the thinning crust apart to form a deep rift. While the rift proceeds across the continent, large earthquakes rattle the region.

Volcanic activity is also prevalent due to the abundance of molten magma rising from the mantle as it nears the surface. Since the crust beneath a rift is only a fraction of its original thickness, magma finds an easy way out. As the crust continues to thin, magma approaches the surface, causing extensive volcanism. A marked increase in volcanic activity during the early stages of many rifts produces vast quantities of lava that flood the landscape. For example, during the rifting of Pangaea into the present continents some 200 million years ago, about 2.7 million square miles, an area nearly the size of Australia, of basalt flooded across the four continents surrounding the Atlantic Basin.

As the rift spreads farther apart and floods with seawater, it eventually becomes a new sea. As the rift continues to widen and deepen, it is replaced by an oceanic spreading-ridge system. This occurs where hot material from the mantle wells up through the rift to form new oceanic crust between the two separated segments of continental crust.

When continental and oceanic plates converge, the denser oceanic plate dives underneath the lighter continental plate, forcing the oceanic plate farther downward. The sedimentary layers of both plates are squeezed like an accordion, swelling the leading edge of the continental crust to create folded mountain belts. The sediments are faulted at or near the surface, where the rocks are brittle, and folded at depths, where the rocks are ductile.

In the deepest parts of the continental crust, where temperatures and pressures are very high, rocks partially melt and become metamorphosed. As the descending plate dives farther under the continent, it reaches depths where the temperatures are extremely high. The upper part of the plate melts to form a silica-rich magma that rises toward the surface because of its greater buoyancy. The magma intrudes the overlying metamorphic and sedimentary layers of the continental crust. There it either forms large granitic bodies or erupts onto the surface.

Magma extruded onto the surface builds volcanic structures, including mountains and broad plateaus. The volcanoes of the Cascade Range came into existence when the Juan de Fuca plate subducted beneath the northwestern United States along the Cascadia subduction zone. As the plate melts while diving into the mantle, it feeds molten rock to magma chambers underlying volcanoes. The subducting plate also has the potential of generating very strong earthquakes, similar to those that plague Chile and Japan.

When two continental plates collide, the topmost layers of the descending oceanic plate are scraped off and plastered against the swollen edge of the continental crust, forming a sedimentary heap called an accretionary wedge (Fig. 17). The submerged crust is underthrusted by additional crustal material.

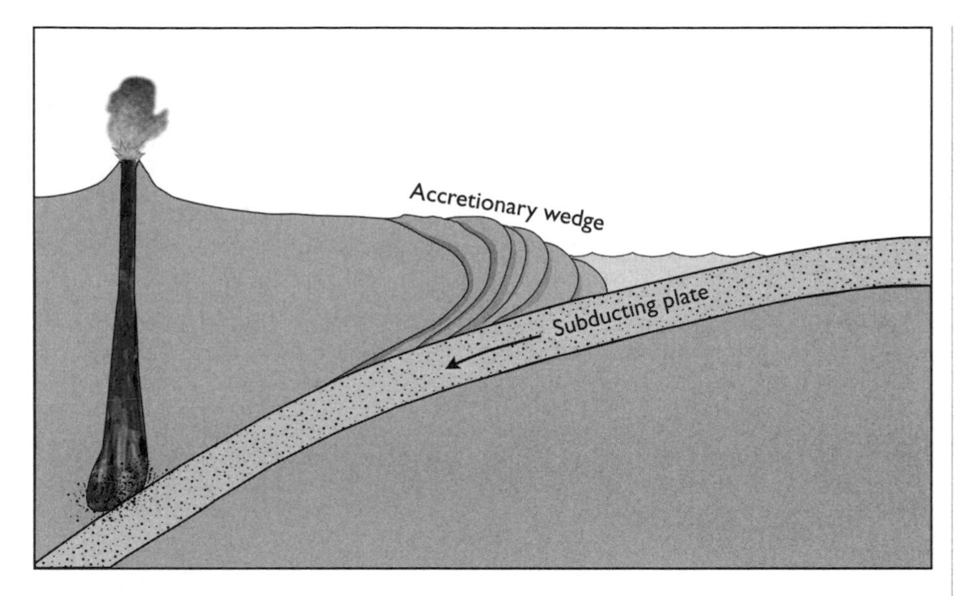

Figure 17 An accretionary wedge is formed by accumulating layers of sediment from a descending oceanic plate as it subducts into the Earth's mantle.

The increased buoyancy raises mountain ranges such as the Himalayas, which arose from the collision of the Indian and Eurasian plates. As a result, the Eurasian plate has shrunk some 1,000 miles since convergence began some 45 million years ago. Additional compression and deformation developed farther inland beyond the line of collision, creating the 3-mile-high Tibetan Plateau with numerous active volcanoes. The strain of raising the world's largest mountain range has resulted in strong deformation accompanied by powerful earthquakes all along the plate.

In submarine interactions, the divergence of lithospheric plates creates new oceanic crust, while convergence destroys old oceanic crust in subduction zones. Subduction of lithospheric plates is most prevalent in the western Pacific, where deep subduction zones are responsible for creating island arcs. Volcanoes of the island arcs are highly spectacular. Their magmas are rich in silica and contain substantial amounts of volatiles, contrasting strongly with the fluid basalts of other volcanoes and spreading ridges. The volcanoes are highly explosive and build steep-sided cinder cones, composed of a composite of cinder and lava. Island arcs are also associated with belts of deep-seated earthquakes several hundred miles below the Earth's surface.

Rifts open under continents as well as ocean basins. The best example is the great East African Rift Valley. It marks the boundary between two tectonic plates, the Nubian plate to the west and the Somalian plate to the east. The rift will eventually widen and flood with seawater to form a new subcontinent similar to Madagascar. This type of rifting is presently taking place in the Red Sea, where Africa and Arabia are diverging. The Gulf of Aden is a young oceanic rift between Africa and Arabia, which have been pulling away from each other for more than 10 million years.

During the rifting process, large earthquakes strike the region as huge blocks of crust drop downward along diverging faults. In addition, volcanoes erupt due to the proximity of the mantle to the surface, which provides a ready supply of magma. A marked increase in volcanism produces vast quantities of basalt lava, which floods onto the continent during the early stages of rifting.

Sometimes, an old extinct rift system, where the spreading activity has ceased, is overrun by a continent. For example, the western edge of the North American continent has overridden the northern part of the now-extinct Pacific rift system, forming California's San Andreas Fault (Fig. 18, 19). A failed rift system beneath the central United States created the New Madrid Fault, which generated three tremendous earthquakes in the winter of 1811–12. The

Figure 18 *The San Andreas Fault in southern California.*

(Photo by R. E. Wallace, courtesy USGS)

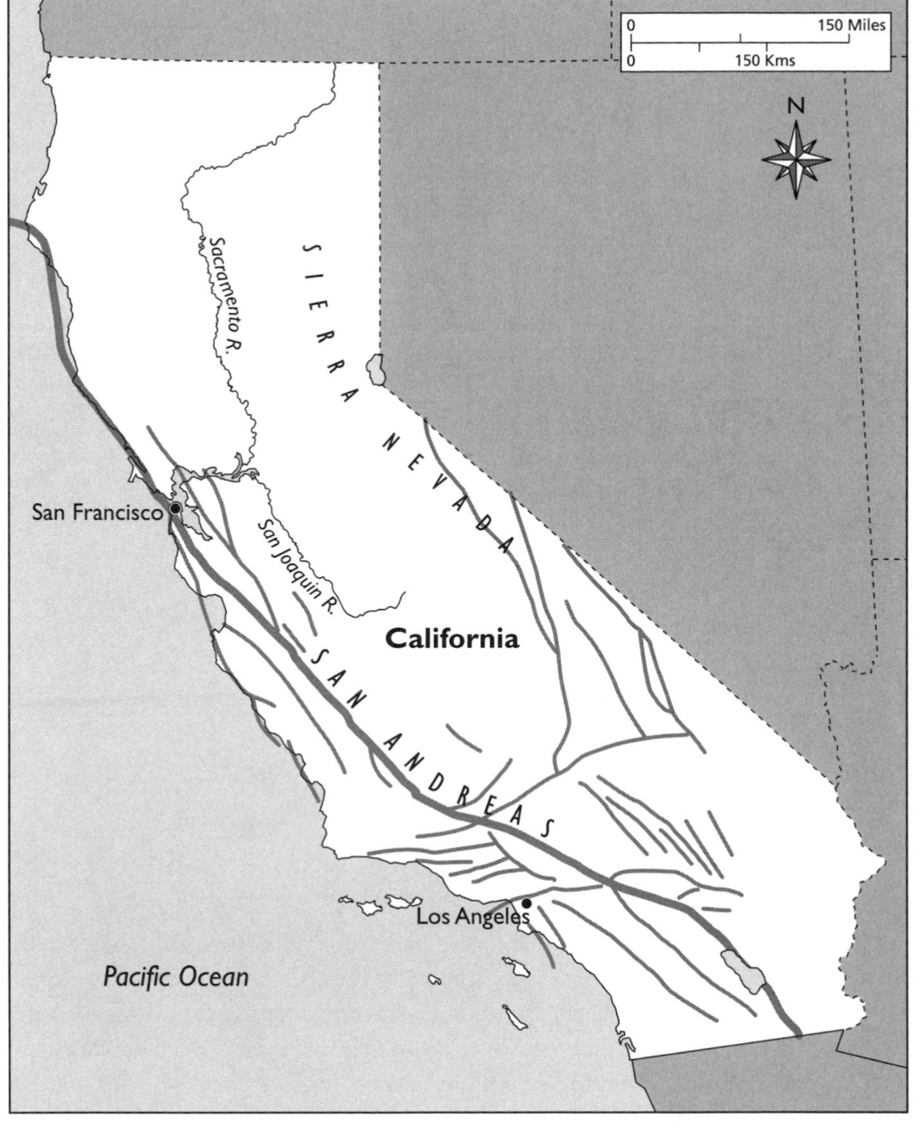

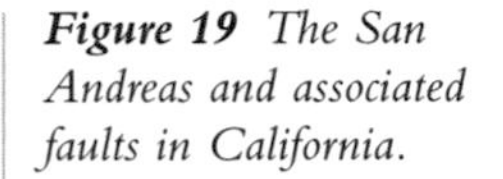

Figure 19 The San Andreas and associated faults in California.

potential of future great earthquakes in both of these regions remains dangerously high.

After discussing the geologic forces that shape our planet, the next two chapters will concentrate on the most destructive forces, namely earthquakes and volcanoes, including their risks, causes, and effects on society.

2

EARTHQUAKES

THE SHAKING OF THE GROUND

This chapter examines the causes and effects of ground shaking by earthquake faults. Earthquakes are by far the most destructive short-term natural forces on Earth. They have plagued civilizations for millennia. Powerful earthquakes with magnitudes greater than 8.0 destroy entire cities, often killing thousands of people with a single massive jolt (Table 2). (Note that the earthquake magnitude scale is logarithmic. An increase of 1 magnitude signifies 10 times the ground motion and a release of 30 times the energy. Only a few earthquakes with a magnitude greater than 9 have been recorded.)

The damage arising from a major earthquake is widespread, changing the landscape for thousands of square miles. Earthquakes often produce tall, steep-banked scarps and cause massive landslides that scar the countryside. Active faults crisscross much of the land surface at plate boundaries near the edges of continents. Half the world's population lives in coastal regions, where they are extremely vulnerable to earthquake destruction.

TABLE 2 THE MOST DESTRUCTIVE EARTHQUAKES

Date (A.D.)	Region	Magnitude	Death Toll
365	Eastern Mediterranean		5,000
478	Antioch, Turkey		30,000
856	Corinth, Greece		45,000
1042	Tabriz, Iran		40,000
1556	Shenshu, China		830,000
1596	Uryu-Jima, Japan		4,000
1737	Calcutta, India		300,000
1755	Lisbon, Portugal		60,000
1757	Concepcion, Chile		5,000
1802	Tokyo, Japan		200,000
1811	New Madrid, Missouri		<1,000
1812	Caracus, Venezuela		10,000
1822	Valparaiso, Chile		10,000
1835	Concepcion, Chile		5,000
1857	Tokyo, Japan		107,000
1866	Peru and Ecuador		25,000
1877	Ecuador		20,000
1883	Dutch Indies		36,000
1891	Mino-Owari, Japan		7,000
1902	Martinique, West Indies		40,000
1902	Guatemala		12,000
1906	San Francisco, California	8.2	3,000
1908	Messina, Sicily	7.5	73,000
1915	Italy		29,000
1920	Kansu, China	8.6	180,000
1923	Tokyo/Yokohama, Japan	8.3	143,000
1927	China	8.6	70,000
1935	Quefta, Pakistan		40,000
1939	Concepcion, Chile		50,000
1939	Erzincan, Turkey	7.9	23,000
1949	Tadzhikstan		12,000

(continues)

TABLE 2 (CONTINUED)

Date (A.D.)	Region	Magnitude	Death Toll
1949	Ecuador		6,000
1953	Greece		3,000
1960	Agadir, Morocco	5.7	12,000
1960	Chile	9.5	6,000
1962	Iran		12,000
1968	Iran		12,000
1970	Peru		67,000
1972	Iran		5,500
1972	Managua, Nicaragua	6.2	12,000
1976	Guatemala	7.5	22,000
1976	Tangshan, China	7.6	240,000
1976	Turkey	7.3	4,000
1978	Eastern Iran		25,000
1980	Southern Italy		45,000
1981	Southeastern Iran		8,000
1982	Northern Yemen		3,000
1985	Mexico City, Mexico	7.8	8,000
1988	Spitak, Armenia	6.9	8,000
1990	Northern Iran		100,000
1995	Kobe, Japan	7.2	5,500
1999	Northern Turkey	7.4	17,000

MAJOR QUAKES

Perhaps the oldest recorded earthquake occurred in China in 1831 B.C. Following a temblor in 1177 B.C., the Chinese began keeping regular records of earthquakes. The first large earthquake, recorded in A.D. 7, destroyed the entire city of Hsien. One of the most destructive earthquakes struck Shenshu in 1556, killing more than 800,000 people and devastating a 500-mile-wide area. The 1920 Kansu earthquake unleashed landslides that took the lives of some 180,000 people. On July 28, 1976, a powerful earthquake in Tangshan in northeast China, about 110 miles east of Beijing (Peking), killed an estimated 240,000 people and left the city in ruins (Fig. 20).

Figure 20 *A crumpled bridge from the July 28, 1976, Tangshan, China, earthquake.*

(Courtesy USGS)

Neighboring Japan, too, has been devastated by earthquakes throughout history. Quakes in Tokyo took the lives of 200,000 people in 1802 and more than 100,000 in 1857. On September 1, 1923, an 8.3 magnitude earthquake struck the Kwanto Plain in central Honshu. The area of total destruction was 90 miles by 50 miles. Huge fissures appeared in the ground, and large landslides permanently altered the landscape. The wreckage was most severe in Tokyo and Yokohama. The triple shocks brought every building of consequence in Yokohama tumbling to the ground. The downtown area of Tokyo was almost completely obliterated (Fig. 21).

Both cities were engulfed by massive firestorms that burned everything in their paths. Nearly three-quarters of the capital city was wiped out. Stirred up by fierce winds, the fires burned for two days, destroying more than 300,000 buildings. Similar whirlwinds in Yokohama swept the flames through every part of the city, and 60,000 buildings burned to the ground. The earthquake destroyed about $3 billion in property and left more than 1 million homeless. When the disaster was over, 140,000 people were counted among the dead.

India has been rattled by earthquakes for ages. On October 11, 1737, a strong earthquake hit Calcutta, killing about 300,000 people. One of the most powerful earthquakes in history struck the Assam region in northeast India on

Figure 21 *The September 1, 1923, earthquake destruction of Main Street, Tokyo, Japan.*
(Courtesy NOAA)

June 12, 1897. The area of total destruction covered about 9,000 square miles, with changes of ground level occurring over large areas. A similar earthquake of 8.7 magnitude hit the region again on August 15, 1950, churning 10,000 square miles of landscape into desolation. Luckily, the area was sparsely inhabited by primitive mountain tribes and the death count was probably low. A 6.4 magnitude earthquake ripped through a large area of Maharashtra State in southwest India on September 30, 1993. The temblor killed 12,000 people (some estimates are as high as 30,000), most of whom were asleep in stone and mud huts that crumbled down upon them. The January 26, 2001, earthquake of 7.9 magnitude in the heavily populated region of Gujarat in western India killed some 30,000 people, mostly when poorly constructed apartment buildings collapsed on them.

To the west, a 6.9 magnitude earthquake of colossal destruction leveled Spitak, Armenia, and other neighboring cities in August 1988. The area lies atop an active thrust fault created by tectonic forces deep in the bowels of the Earth. The poorly designed concrete buildings could not withstand the earthquake's terrible punishment as the ground suddenly lifted as much as 6 feet. At least 25,000 people lost their lives when buildings crashed down onto them. Nearby, in northern Iran, a powerful earthquake in June 1990 triggered massive landslides that killed 100,000 people and left half a million homeless.

Farther to the west in Turkey, the ancient city of Antioch, now Antakya, was built partly on soft ground and has suffered from earthquakes since it was founded. In A.D. 115, the city was almost totally destroyed. Because of its strategic military position in southern Turkey near the border with Syria, it was rebuilt on the same site. In A.D. 458, Antioch was destroyed again by another earthquake and rebuilt on the same spot even though reconstruction here was thought to be unwise. As predicted, the part of the city rebuilt on the worst ground near the river was destroyed a generation later when another earthquake hit and 30,000 lives were lost. Again, the city was rebuilt on the same site and became an important religious center only to be destroyed a final time by the Persians in A.D. 540.

Continuing westward, a tremendous force was unleashed against Lisbon, Portugal, on November 1, 1755, causing buildings to crumble and fall to the ground (Fig. 22). The shock was felt throughout Portugal, Spain, and other parts of the world as far away as the United States. Many areas along the coast of Portugal came to rest at new levels. A second earthquake struck 20 minutes later. In the renewed shaking, the stone quay gave way and sank into the river, taking people who sought safety on the riverfront along with it. A 20-foot tsunami generated by the undersea quake swept through the harbor, destroy-

Figure 22 *Illustration of the collapsing city during the November 1, 1755, Lisbon, Portugal, earthquake.*

(Courtesy USGS)

ing bridges and overturning ships. A vast inferno that burned for several days enveloped the city, turning it to ashes. The earthquake completely leveled the city and killed 60,000 of its inhabitants. The powerful quake might have triggered sympathetic tremors in North Africa, causing heavy damage there as well.

Just before noon on June 7, 1692, Port Royal on the island of Jamaica began to shake in three separate shocks. The ground rose and fell in waves, cracking open and closing again, swallowing people and crushing them to death. The earthquake was accompanied by a great roar. The whole north end of town slowly slid into the sea. Along the waterfront, buildings toppled over and sank beneath the waves. Ships in the harbor capsized by the turbulence of the sea. Some 2,000 people disappeared, and two-thirds of the city was destroyed. In its place, the new port of Kingston was built only to be destroyed by fire following an earthquake in 1907.

New England has had a long history of earthquakes. Some of the first explorers were startled when they felt the shaking and rumbling of the earth. The Native Americans before them knew the experience quite well. The Pilgrims who landed at Plymouth felt their first major quake in 1638, which rattled buildings and instilled panic throughout the countryside. In 1727, a temblor shook the East Coast from Maine to Delaware. The shaking in the town of Newbury, Massachusetts, leveled many chimneys and stone walls. An even stronger shock rocked eastern Massachusetts in 1755. It imparted major damage from Cape Ann to Boston, where streets were so cluttered with debris from crumbling buildings they were practically impassable. Another historic earthquake struck the town of Moodus, Connecticut, in 1791, causing some minor building damage. Interestingly, many legends were attributed to the "Moodus noises," which were actually caused by swarms of small earthquakes very near the surface.

Three of the greatest earthquakes to strike the continental United States shook New Madrid, in southeastern Missouri, on December 16, 1811, January 23, 1812, and February 7, 1812. The town, located on the banks of the Mississippi River, was the largest settlement in the region. The quakes thoroughly destroyed nearly all buildings as the ground rose and fell. Trees snapped in two, the earth split open and formed deep fissures, and massive landslides slid down bluffs and low hills (Fig. 23). Thousands of broken trees fell into the river. Sandbars and islands disappeared. Great waves were created on the river, overturning many boats and washing others high upon the shore. The earthquakes changed the path of the river, which wandered far to the west of its normal course. The downdropped crust formed deep lakes.

The shocks were felt in many parts of the country. The vibrations alarmed the inhabitants of Chicago and Detroit, who experienced slight tremors. The trembling woke people in Washington, D.C., and rang church

Figure 23 *Landslide fissures in Chickasaw Bluff, east of Reelfoot Lake, Tennessee, produced by the 1811 New Madrid earthquake.* (Photo courtesy USGS)

bells in Boston more than 1,000 miles away. Pendulum clocks stopped throughout the eastern part of the country. Near New Madrid, minor aftershocks were felt for the next two years. If similar quakes hit the same area today, the damage would be colossal because the region now contains many large cities, with a total population exceeding 15 million.

The most powerful earthquake in California's recent history ruptured the southern end of the San Andreas Fault near Los Angeles in 1857. The hamlet of Lone Pine in the Owens Valley, east of the Sierra Nevada, was destroyed on March 26, 1872, by another large earthquake. It opened a deep fissure along a 100-mile line in the Owens Valley. At least 30 people died when their fragile adobe huts collapsed onto them. More than 1,000 aftershocks ran through the area during the next three days.

On April 18, 1906, the northern end of the San Andreas Fault ruptured near San Francisco. An area of 4 square miles, covering 75 percent of the city, was destroyed (Fig. 24). The damage was greatest in the low-lying business section. Almost all the buildings in the downtown area were destroyed or structurally weakened. The ground subsided under some buildings, causing them to collapse. Houses reeled and tumbled over. In moments, streets were piled high with debris. Fractured gas mains were ignited by overturned stoves and electric sparks from overhead wires, turning the city into an inferno. Those

Figure 24 *Damage to buildings viewed from the corner of Geary and Mason Streets from the April 18, 1906, San Francisco earthquake.*

(Photo by W. C. Mendenhal, courtesy USGS)

buildings that managed to survive the earthquake were utterly destroyed by the subsequent fire. The death toll reached 3,000 people, and 300,000 were left homeless.

The town of Santa Rosa, 50 miles to the north, was nearly totally wrecked by the 8.2 magnitude earthquake. South of Cape Fortunas, a hill slid altogether into the sea and created a new cape. The road between Point Reyes Station and Inverness was offset horizontally 21 feet where it crossed the San Andreas Fault. Trees were uprooted, and fissures and springs appeared in many districts. A repeat performance played out in San Francisco on October 17, 1989, although the damage and the loss of life was not nearly as great. Had this one been as powerful as the 1906 earthquake, damages could have reached $40 billion, accompanied by tremendous casualties.

On Good Friday, March 27, 1964, the largest recorded earthquake to hit the North American continent devastated Anchorage, Alaska, and surrounding areas. The 9.2 magnitude quake caused destruction over an area of 50,000 square miles and was felt throughout an area of half a million square miles. Anchorage bore the brunt of the earthquake, and 30 city blocks were destroyed when the ground subsided several feet. Along a section of Fourth Avenue, one whole city block gave way. This caused a row of cafes and pawnshops, along with cars and the street itself, to drop down to the basement level (Fig. 25).

Huge fissures opened in the outlying areas, and the greatest crustal deformation ever known took place. Landslides caused much of the damage. Entire port facilities slid bodily into the sea. The port of Seward experienced a remarkable landslide as the seashore slid into the bay. At Resurrection Bay, the water was disturbed by incredible turbulence as it rebounded from one side of the narrow inlet to the other. A 30-foot-high tsunami generated by the undersea earthquake destroyed coastal villages around the Gulf of Alaska, killing 107 people. Kodiak Island was heavily damaged, and most of the fishing fleet was destroyed when the tsunami carried many vessels inland.

Latin America has had more than its fair share of destructive earthquakes. In 1730 and again in 1751, Chile was shaken by enormous earthquakes. Spectacular ground movements were recorded at Riocamba during the earthquake of 1797. A major earthquake struck Valdivia and Concepción in 1835. Concepción was destroyed for a sixth time and Valdivia, Puerto

Figure 25 *The collapse of Fourth Avenue in Anchorage from the March 27, 1964, Alaskan earthquake.* (Courtesy U.S. Army and USGS)

Montt, and other ports were wrecked by an earthquake of 9.5 magnitude on May 22, 1960. Giant tsunamis raced across the coast, landslides wasted the countryside, and two dormant volcanoes came to life in an area of 90,000 square miles. Some 50,000 homes were destroyed, and 5,700 people lost their lives. The December 23, 1972, Managua, Nicaragua, earthquake destroyed 36 city blocks and killed some 10,000 people. The 1976 Guatemala City earthquake took the lives of about 23,000 people, injured 77,000, and left 1 million homeless (Fig. 26).

On September 19, 1985, an 7.8 magnitude earthquake and a 7.6 magnitude aftershock toppled buildings in Mexico City. It was called the gravest disaster in Mexico's history. The earthquake and aftershock killed upward of 10,000 people, and left some 100,000 homeless. So violent was the earthquake that buildings trembled in Texas and water sloshed in Colorado swimming pools. Buildings in downtown Mexico City vibrated wildly. Their walls and girders groaned from the extreme stress. Metal lampposts swayed and bent like rubber in the shuddering streets. Telephone and electric wires snapped, windows shattered, and huge chunks of concrete broke off buildings and smashed

Figure 26 *Sustained earthquake damage in the town of Joyabaj, Department of Quiche, Guatemala, in 1976.* (Courtesy USGS)

Figure 27 *Buildings damaged by the September 19, 1985, Mexico City earthquake.* (Photo by M. Celebi, courtesy USGS)

to the pavement below. Some 400 buildings crumbled in the quakes, and 700 others were severely damaged (Fig. 27).

AREAS AT RISK

Most earthquakes originate at plate boundaries, where lithospheric plates are converging, diverging, or sliding past each other. The most powerful earthquakes are associated with plate subduction, where one plate thrusts under another. The area of greatest seismicity corresponds with deep trenches and volcanic island arcs, which mark the convergence between oceanic and continental plates. Earthquakes also coincide with midocean rift systems, which extend thousands of miles through the world's oceans. Earthquakes associated with terrestrial rift zones, such as the 3,600-mile East African Rift, are highly destructive.

Most earthquakes concentrate in broad zones that wind around the globe (Fig. 28). The greatest amount of seismic energy is released along the rim of the Pacific Ocean, known as the circum-Pacific belt. This is a band of subduction zones that flank the Pacific Basin. It coincides with the Ring of

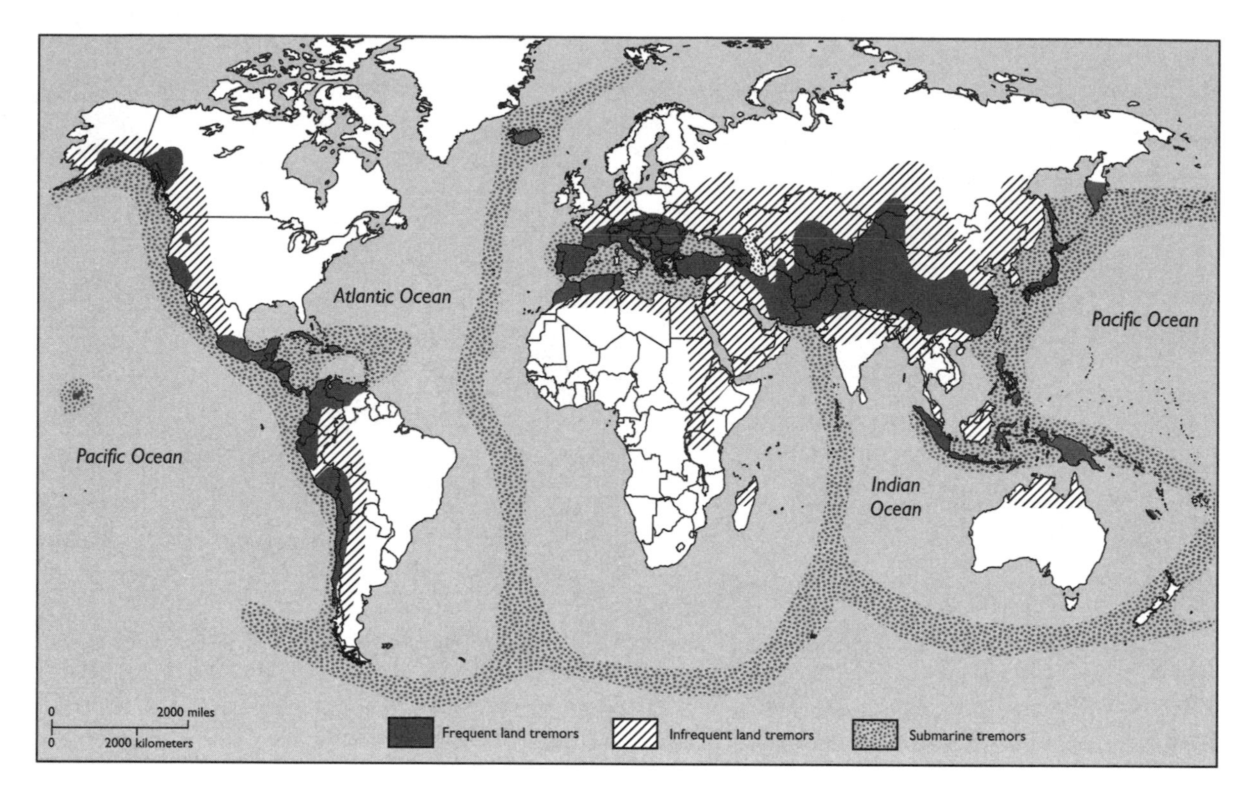

Figure 28 Earthquake belts associated with plate boundaries, where most seismic activity occurs.

Fire, which explains why the Pacific Rim also contains most of the world's active volcanoes. In the western Pacific, the circum-Pacific belt encompasses the volcanic island arcs that fringe the subduction zones, producing some of the largest earthquakes on Earth.

When beginning at New Zealand, the circum-Pacific belt runs northward over the islands of Tonga, Samoa, Fiji, the Loyalty Islands, the New Hebrides, and the Solomons. The belt continues westward over New Britain, New Guinea, and the Moluccas islands. One segment traverses toward the west over Indonesia. The principal arm continues northward over the Philippines, where a large fault zone spans the entire length of the islands. The band of earthquakes proceeds on to Taiwan. There, on September 28, 1999, in Taipei, the largest earthquake in over a century, registering 7.6 magnitude, killed more than 2,200 people and left 100,000 homeless. The Japanese archipelago has been hard hit by major quakes. The January 17, 1995 Kobe earthquake of 7.2 magnitude killed more than 5,500 people and caused over $100 billion in property damage.

The belt continues northward and follows the seismic arc across the top of the Pacific, including the Kuril Islands, the Kamchatka Peninsula, and the Aleutian Islands. The belt travels along the Aleutian Trench, responsible for many great Alaskan earthquakes, the Cascadia subduction zone, which has shaken the Pacific Northwest in prehistoric times, and the San Andreas Fault that rattles California (Fig. 29).

The belt runs down the Andes Mountains of Central and South America, known for many of the largest and most destructive earthquakes on record. During the last century, nearly two dozen earthquakes with magnitudes of 7.5 or greater have devastated these areas. An immense subduction zone just off the coast threatens the entire western seaboard of South America. The lithospheric plate on which the South American continent rides is forcing the Nazca plate to buckle under, building up great tensions deep within the crust. While some rocks are forced downward, others are pushed to the surface to raise the Andes Mountains, the fastest-growing mountain range on Earth. The resulting forces are building great stresses into the entire region. When the strain becomes too great, earthquakes roll across the countryside.

A second major seismic zone runs through the folded mountain regions that flank the Mediterranean Sea, well known throughout the annals for its destructive earthquakes. The eastern Mediterranean region is a jumble of colliding plates, providing highly unstable ground. The Near East is extremely unstable, attested to by the many earthquakes reported in biblical times. Since

Figure 29 *A view northwest toward San Francisco, California, along the San Andreas Fault.*

(Photo by R. E. Wallace, courtesy USGS)

the latter part of the second millennium B.C., tablets found in Iraq, Syria, and other towns on the Mediterranean coast have told of local rulers being unable to pay taxes after earthquakes destroyed their cities. Roman tax records indicate that numerous towns received financial aid from the government to help repair earthquake damage.

Earthquakes have often ravaged the remaining regions surrounding the Mediterranean since the dawn of civilization. One of the most disastrous earthquakes on record occurred in the region on July 21, A.D. 365. It affected an area of about a million square miles in the eastern Mediterranean, encompassing Italy, Greece, Palestine, and North Africa. The earthquake leveled coastal towns, and a gigantic tsunami destroyed the Egyptian port of Alexandria, drowning 5,000 people.

The ancient city of Curium in southern Cyprus was totally leveled by the earthquake. For centuries, a part of classic Roman civilization was buried in the rubble. Archaeologists excavating in the area have uncovered well-preserved artifacts along with bones of humans and animals. Analysis of walls and objects falling to the ground indicate that the earthquake was of extreme intensity, causing near-total destruction. People and animals were immediately trapped in buildings as the walls literally shattered.

The belt continues through Iran and past the Himalaya Mountains into China. At the eastern end of the Himalayas lies possibly the most earthquake-prone area in the world. An enormous 2,500-mile-long seismic zone stretches across Tibet and much of China. In the last century, more than a dozen earthquakes with magnitudes of 8.0 or greater have struck the region. Westward, the Hindu Kush Range of northern Afghanistan is the site of many devastating earthquakes. During the 20th century, the region has witnessed three great earthquakes of magnitude 8.0 or larger. This is a highly active earthquake belt, with some 2,000 minor quakes occurring annually.

The seismic zone continues through the Caucasus Mountains on to Turkey. In 1939, a 7.9 magnitude earthquake in eastern Turkey killed some 30,000 people. The August 17, 1999, earthquake of 7.4 magnitude occurred along a fault system similar to the San Andreas. It killed more than 17,000 people in the industrial heartland. The quake occurred near the western end of the North Anatolian Fault, a 750-mile-long tear in the crust extending across the northern part of the country. Unrest along the fault stems from the slow-motion collision between Arabia and Eurasia.

Earthquakes also occur in stable zones that comprise the strong rocks of the Precambrian shields in the interiors of the continents. The stable zones include Scandinavia, Greenland, eastern Canada, parts of northwestern Siberia and Russia, Arabia, the lower portions of the Indian subcontinent, the Indochina peninsula, almost all of South America except the Andean mountain region, much of Australia, and the whole of Africa except the Great Rift

Valley and northwestern Africa. Earthquakes in these regions might have been triggered by the weakening of the crust by compressive forces originating at plate edges. The crust might also have been weakened by previous tectonic activity involving extinct or failed rift systems similar to the New Madrid Fault, which, according to geologic history, is poised for another major quake.

EARTHQUAKE FAULTS

The mechanics of earthquakes were not well understood until after the great 1906 San Francisco quake along the northern San Andreas Fault. For hundreds of miles, roads and fences crossing the fault were offset by as much as 21 feet (Fig. 30). The San Andreas runs for 650 miles from the Mexican border through the western edge of California and plunges into the Pacific Ocean near Cape Mendocino in the northern part of the state. It represents the

Figure 30 *A fence in Marin County offset 8.5 feet by the San Andreas Fault during the April 18, 1906, San Francisco, California, earthquake.*

(Photo by G. K. Gilbert, courtesy USGS)

boundary between the Pacific and North American plates. The segment of California west of the San Andreas is sliding past the North American continent in a northwestward direction about 2 inches per year. The relative movement of the two plates is right lateral or dextral because an observer on either side of the fault would notice the other block moving to the right. During the San Francisco earthquake, the Pacific plate suddenly slipped several feet relative to the North American plate.

During the 50 years since the last major earthquake that struck southern California in 1857, the rocks along the San Andreas Fault were bending and storing elastic energy as a stick stores elastic energy when it is bent. Eventually, the forces holding the rocks together lost their strength and slippage occurred at the weakest point. The displacement exerted strain farther along the fault until most of the built-up strain was released. Like a stick bent to its maximum, it snapped.

The sudden slippage allows the deformed rock to rebound. As the rock elastically returns to its original shape, it produces vibrations called seismic waves, which are similar to sound waves. The seismic waves radiate in all directions, like ripples generated by tossing a pebble into a quiet pond. The rocks do not always rebound immediately, however, but might take days or even years, resulting in aseismic or nonearthquake slippage. This condition generally occurs along the midsection of the San Andreas Fault, which produces relatively mild earthquakes because the two sides of the fault generally slide smoothly past each other. Conversely, at the southern end of the fault, responsible for the massive 1857 quake, and at the northern end of the fault, known as the big bend and responsible for the 1906 San Francisco earthquake, the ends of the fault tend to snag. When the locked portions of the fault attempt to tear free, powerful earthquakes rumble across the landscape.

An unusual set of circumstances caused by lateral slippage at the southern end of the San Andreas Fault combined with plate subduction at the northern end resulted in a major earthquake of 6.1 magnitude in southern California near Joshua Tree on April 22, 1992, and two more to the north three days later. The largest quake to hit California in 30 years was the June 28, 1992, Landers earthquake of 7.5 magnitude. Three days later, a 6.5 magnitude earthquake shook the town of Big Bear, 20 miles to the northwest. Luckily, the strongest seismic waves raced farther northward into the sparsely populated Mojave Desert rather than heading into the crowded urban areas to the west.

The tremors might herald a great quake, the "Big One," as Californians say. They could have a direct effect on a nearby segment of the San Andreas Fault by shifting stresses in the crust, which make conditions easier for an earthquake to occur along the fault. The quakes might also signal the birth of a new competing fault parallel to the San Andreas.

Faults are also displaced in a vertical direction, with one side positioned higher with respect to the other side (Fig. 31). When the crust is pulled apart, a fault block slides downward along an inclined plane. This results in a gravity or normal fault, a historical misnomer because this was once thought to be how faults normally occur. Actually, most faults arise from compressional forces that push fault blocks upward along an inclined plane. These are called reverse faults because they are the opposite of gravity faults. If the reverse fault plane is nearly flat and the movement is mostly horizontal, a thrust fault results. A thrust fault forms when a compressed plate shears, causing one section to be lifted over another for long distances.

Thrust faults are not always exposed on the surface. Some thrust faults associated with the San Andreas Fault lie deep underground, where stresses along the fault increase with depth. Concealed faults beneath Los Angeles are called blind thrust faults because they remain out of sight, covered over by thick layers of sediment. When a deep fault fails to break the surface during a major earthquake, it might be due to a blind, low-angle thrust fault.

A thrust fault lying about 6 miles beneath the surface produced a 6.7 magnitude earthquake at Coalinga, California, on May 2, 1983, that nearly leveled the town. The earthquake was blamed on a thrust fault because of the lack of ground rupture, which should have occurred with any temblor greater than 6.0. The unusual earthquake had no recognizable foreshocks that would have provided advance warning. The earthquake that struck Whittier, California, on October 1, 1987, was a little less than 6.0 magnitude. Although the fault did not

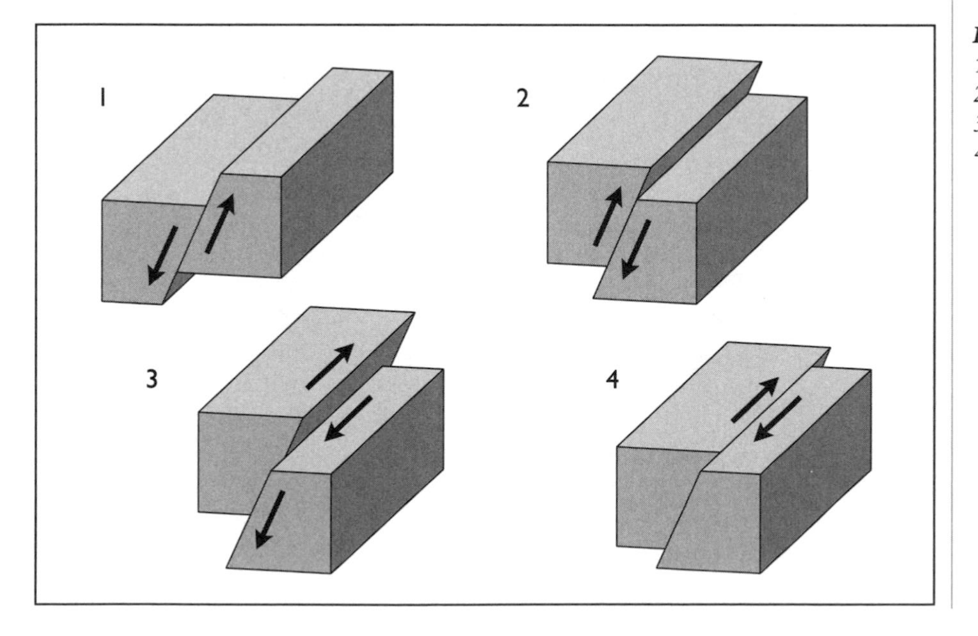

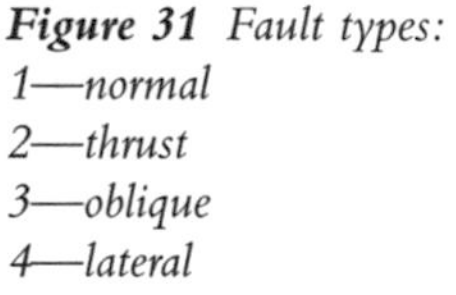
Figure 31 *Fault types:*
1—normal
2—thrust
3—oblique
4—lateral

Figure 32 *Serious damage from the 1980 El Asnam, Algeria, earthquake.*

(Photo courtesy NOAA)

rupture the surface, damage was severe. The hills outside town grew almost 2 inches. The January 17, 1994, Northridge earthquake of magnitude 6.7 occurred on a thrust fault that caused severe damage in the Los Angeles area, killing 63 people, leaving 10,000 homeless, and costing upward of $13 billion.

Thrust faults associated with the San Andreas Fault system might be expressed on the surface as a series of active folds that continuously uplift California's Coast Ranges. Earthquakes upraise large folds called anticlines that arise from a series of discrete tectonic events over millions of years. Most of these earthquakes occur under young anticlines less than a few million years old because folded strata are the geologic product of successive earthquakes resulting from compressional forces during plate collisions. Unlike earthquakes that break along faults, those associated with folds do not rupture the Earth's surface.

Many of the world's major fold belts that raised mountain ranges, such as those bordering the Mediterranean Sea, are earthquake prone. For example, an anticline associated with the fault responsible for the 1980 El Asnam, Algeria, earthquake (Fig. 32) was uplifted more than 15 feet. During the 20th century, large fold earthquakes have taken place in Japan, Argentina, New Zealand, Iran, and Pakistan.

Thrust faults can cause considerably more damage than lateral faults with equal measures of magnitude. Lateral faults cause buildings to sway back and forth, allowing their flexible frames to absorb most of the force. Some buildings sway so fiercely during an earthquake, however, they seriously injure occupants, as furniture and other objects (including people) are hurled against walls. Thrust faults, in contrast, suddenly raise and drop buildings inches or feet at a time, creating tremendous forces that topple even the best-designed structures. The strongest shaking occurs on the hanging wall portion of the fault, the land that rises during an earthquake. A thrust fault responsible for the 1988 Armenian earthquake caused substantial damage and the loss of 25,000 lives.

Some faults are a combination of horizontal and vertical motions. They consist of complicated diagonal movements that form complex fault systems called oblique or scissors faults. The great Uinta Fault on the north side of the Uinta Mountains in Utah is an example of such a fault. The October 17, 1989, Loma Prieta earthquake (Fig. 33) ruptured a 25-mile-long segment of the San Andreas Fault. The faulting propagated upward along a dipping plane, resulting in a right oblique reverse fault. The earthquake raised the southwest side of the fault more than 3 feet, contributing to the continued growth of the nearby Santa Cruz Mountains.

Both the 1906 San Francisco and the Loma Prieta earthquakes took place on a segment of the San Andreas Fault that runs through the Santa Cruz

Figure 33 *Buildings damaged in the Marina district, San Francisco, from the October 17, 1989, Loma Prieta, California, earthquake.*

(Photo by G. Plafker, courtesy USGS)

Mountains. The major difference between the two is that most of the motion of the earlier quake was horizontal, whereas the latter quake occurred along a tilted surface that forced the southwest side of the fault to ride over the northeast side. Because of the subsurface geology of the region, the shaking of the San Francisco area reached double the level expected for an earthquake of this size.

The rest of the nation is crisscrossed by numerous faults generally associated with mountains, and 39 states lie in regions classified as having moderate-to-major earthquake risk (Fig. 34). The Basin and Range Province in Oregon, Nevada, western Utah, southeastern California, and southern Arizona and New Mexico contains several fault-block mountains that are prone to earthquakes. The upper Mississippi and Ohio River valleys suffer frequent earthquakes. The northeast-trending New Madrid and associated faults are responsible for major earthquakes and many tremors. Along the eastern seaboard, major earthquakes have hit Boston, New York, Charleston, and other areas. Eastern North Amer-

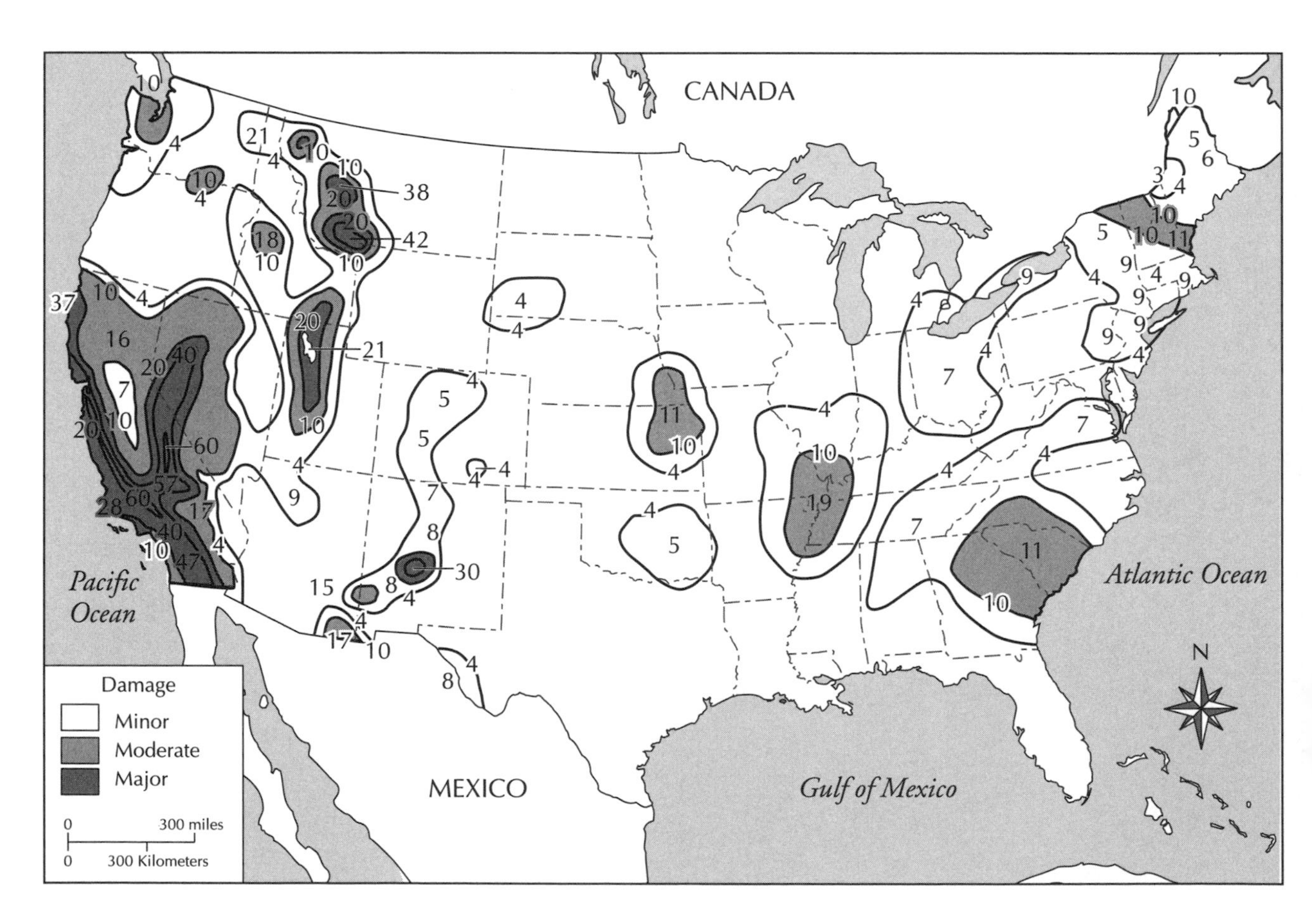

***Figure 34** Earthquake risk areas.*

ica has had more than a dozen moderate-to-large earthquakes since colonial times. The region has been comparably quiet over the years, spurring fears that a major earthquake might be looming just over the horizon.

EARTHQUAKE CAUSES

Why seismic energy releases violently in some cases and not in others is not fully understood. An earthquake's magnitude is proportional to the length and depth of the rupture created by slipping plates. Generally, the deeper and longer the fault is, the larger the earthquake. The great 1960 Chilean earthquake, the largest ever recorded, was generated along a 600-mile-long rupture through the South Chilean subduction zone, which might have slipped all at once. During the 1906 San Francisco earthquake, a 260-mile section of the San Andreas Fault ruptured. Other processes that affect earthquake magnitude include the frictional strength of the fault, the drop in stress across the fault, and the speed of the rupture as it traverses over the fault. A break along a fault can travel at speeds of up to a mile a second.

The Earth's crust is constantly readjusting itself, producing vertical and horizontal offsets on the surface. These movements are associated with large fracture zones in the crust. The greatest earthquakes have offsets of several tens of feet occurring in a matter of seconds. Most faults are associated with plate boundaries, and most earthquakes are generated in zones where huge plates collide or shear past each other. If the plates hang up in so-called "stuck spots" known as asperites, the sudden release generates tremendous seismic energy. The interaction of plates causes rocks to strain and deform. If deformation takes place near the surface, major earthquakes result. Earthquakes also occur during volcanic eruptions, but they are relatively mild compared with those caused by faulting.

Thousands of earthquakes strike yearly. Fortunately, only a few are powerful enough to be destructive. During the 20th century, the world average was about 18 major earthquakes of magnitudes 7.0 or greater per year. For great earthquakes with magnitudes above 8.0, the century's average was 10 per decade. The degree of damage does not depend on magnitude alone, which is proportional to the length and depth of the rupture. It is also influenced by the geology of the region. Earthquakes occurring in strong rocks, such as those in continental interiors, are more destructive at equal magnitudes than those occurring in the fractured rock at plate margins. This is why earthquakes in the eastern United States influence a wider area than earthquakes in the West. The August 31, 1886, Charleston, South Carolina, earthquake (Fig. 35), which killed 110 people locally, cracked walls in Chicago 750 miles away and was felt in Boston, Milwaukee, and New Orleans.

Figure 35 *Wreckage from the August 31, 1886, Charleston, South Carolina, earthquake.*
(Photo by J. K. Hillers, courtesy USGS)

The longer the time since the last big shock on a major fault, the greater is the earthquake hazard. This is known as the seismic gap hypothesis, which holds that the earthquake hazard along faults is low immediately following a large earthquake and increases with time. Because much time is needed for strain to build up again, only long-dormant faults are prime hazards. Although the theory applies only to large earthquakes, a moderate quake could strike on the same fault without warning, making predictions very tenuous.

The earthquake hazard depends on the size of the temblor and on the geology of the region. Most faults appear to have a characteristic earthquake that recurs in like manner. Some areas might experience similar earthquakes of 7.0 magnitude, whereas other regions might be prone to great earthquakes of 8 or 9 magnitude. However, larger quakes do not follow the same patterns set by smaller ones, making their prediction extremely difficult. Earthquakes are likely to strike where they have occurred before. Once a zone becomes seismically active, earthquakes continue until, for unknown reasons, they cease. Then a relatively long interval passes before another great one comes along.

EARTHQUAKE DAMAGE

Earthquakes are the most highly destructive geologic forces. The damage arising from a major temblor is widespread, altering the landscape for thousands of square miles. The Earth's crust is constantly readjusting itself. This results in vertical and horizontal displacements on the surface associated with fracture zones in the crust. Large earthquakes can produce offsets of several tens of feet in only a few tens of seconds. The rupturing faults can also communicate with other long-distant faults, causing earthquakes up to thousands of miles away.

Besides the destruction of buildings and other structures, earthquakes alter the landscape by producing deep fissures and tall scarps (Fig. 36) and by

Figure 36 *The Red Canyon fault scarp from the August 1959 Hebgen Lake earthquake is about 14 feet high, Gallatin County, Montana.*

(Photo by I. J. Witkind, courtesy USGS)

causing massive landslides that scar the terrain. The greatest deformation occurs near thrust faults. This is where one block overrides another, especially on the hanging wall, which rises during an earthquake. Active faults, responsible for scarps, rifts, and mountain ranges, crisscross much of the land surface at plate boundaries on the edges of continents and in the continental interiors underlain by old rifts.

Ancient civilizations living in earthquake-prone regions protected themselves from the ravages of quakes by constructing simple dwellings that could withstand violent shaking. Today, however, as accommodations have become more sophisticated with complex construction materials, earthquake damage has become a serious and expensive problem (Fig. 37). Most large urban centers are a combination of old and new buildings, often with modern structures blending in with earlier architecture, whose foundations have weakened with time.

Many areas, such as the Pacific Northwest—which lies along a subduction zone and has been devastated by earthquakes in the distant past, were not built to survive severe ground shaking and are totally unprepared for a major quake. When the February 28, 2001, earthquake of 6.8 magnitude struck the Seattle, Washington, area, it cost more than $2 billion in damages. About 1,000 years ago, a huge earthquake struck the same region. The ground shook with such fury that avalanches and landslides tumbled from the Olympic Mountains and buried areas that are now densely populated. The earthquake also

Figure 37 *Collapse of the Hotel Terminal caused by the failure of reinforced concrete columns during the 1976 Guatemala City, Guatemala, earthquake.*

(Courtesy USGS)

triggered a great tsunami that washed the shores of the Puget Sound. If such an earthquake struck today, the damage to property and the loss of life would be monumental.

The type of construction determines how well a structure survives an earthquake. Lightweight, steel-framed buildings with strength combined with flexibility and also reinforced-concrete buildings with few window and door openings that tend to weaken the structure generally suffer little damage during an earthquake. A building's ability to withstand a major earthquake depends not only on its design, the type of materials used, and the quality of workmanship but also on the type of ground it sits on, its orientation with respect to the shock wave, and the nature of the shock wave.

Designing a structure to withstand a short, sharp, high-frequency shock wave lasting only a few seconds is comparatively easy. Buildings of two to four stories are most vulnerable to this type of shock wave, whereas taller buildings might escape unscathed. Designing a structure to withstand a longer, lower-frequency shock wave lasting several tens of seconds is much more difficult. Multistory buildings are most vulnerable to this type of shock wave, whereas lower buildings remain practically untouched. The longer the earthquake's duration, the more that tall buildings will resonate, causing them to rock back and forth violently.

Even if buildings can withstand an earthquake, they are still vulnerable to foundation failure causing buildings to topple over when the ground gives way beneath them. Severe shock waves can make soils settle or liquefy, and they lose their ability to support structures (Fig. 38). The building site greatly affects how much movement a structure experiences. Generally, structures built on bedrock are damaged less severely than those built on less consolidated, easily deformed materials such as natural and artificial fills. The type of ground that supports a structure affects the amount of movement because soft sediments generally absorb high-frequency vibrations and amplify low-frequency vibrations, which do the most damage.

The length of time an earthquake is in motion greatly affects the amount of damage to buildings. Generally, the longer the earthquake interval is, the more severe the damage. Other factors that determine the degree of earthquake destruction include the type of seismic waves involved. As the energy released by an earthquake travels along the surface, it causes the ground to vibrate in a complex manner, moving up and down as well as from side to side.

Most buildings can handle the vertical motions because they are built against the force of gravity. However, the largest ground motions are usually horizontal, which cause buildings to sway back and forth. If the structure's resonance frequency corresponds to that of the earthquake, it could sway wildly and cause considerable damage. Aftershocks caused by readjustments in rocks

***Figure 38** Highway 1 bridge destroyed by liquefaction of river deposits at Struve Slough during the October 17, 1989, Loma Prieta, California, earthquake.*
(Photo by G. Plafker, courtesy USGS)

following the main event can be just as destructive and finish off what the earthquake started. Further damage results from fires set in broken gas lines and other flammable materials that burn out of control, with fire fighting efforts hampered by broken water mains.

The size of the geographic area influenced by shock waves depends on the magnitude of the earthquake and the rate at which the amplitudes of the seismic waves diminish with distance. Some types of ground transmit seismic energy more effectively than others. For a given magnitude, seismic waves extend over a much wider area in the eastern United States than in the West, which indicates a substantial difference in the crustal composition and structure of the two regions. The East is composed of older sedimentary rock, whereas the West is composed of fairly young igneous and sedimentary rocks fractured by faults.

TSUNAMIS

Undersea earthquakes that vertically displace the ocean floor produce tsunamis, the Japanese word for "harbor waves," so-named because of their common occurrence in that country. The energy of an undersea earthquake transforms into wave energy proportional to its intensity. The earthquake sets up ripples on the ocean similar to those formed by tossing a rock into water. In the open ocean, the wave crests are up to 300 miles long and usually less than 3 feet high, with a distance between crests of 60 to 120 miles. This gives the tsunamis very gentle slopes that pass practically unnoticed by ships or aircraft.

A tsunami extends thousands of feet to the ocean floor and travels at speeds of 300 to 600 miles per hour. Generally, the deeper the water and the longer the wave, the faster the tsunami travels. When the tsunami touches bottom upon entering shallow coastal waters, such as a harbor or narrow inlet, its speed diminishes abruptly to about 100 miles per hour. The sudden breaking action causes seawater to pile up. The wave height is magnified tremendously as waves overtake one another, decreasing the distance between them in a process called shoaling. Tsunamis have been known to grow into a towering wall of water up to 200 feet high, although most are only a few tens of feet high. The destructive power of the wave is immense, causing considerable damage as it crashes to shore. Buildings are crushed with ease, and ships are often carried well inland (Fig. 39).

The Pacific Ocean is responsible for 90 percent of all tsunamis in the world, and 85 percent of those are the products of undersea earthquakes. Between 1992 and 1996, 17 tsunami attacks around the Pacific killed some

Figure 39 *Tsunamis washed many vessels into the heart of Kodiak from the March 27, 1964, Alaska, earthquake.*

(Courtesy USGS)

1,700 people. The Hawaiian Islands are in the paths of many damaging tsunamis. Since 1895, 12 such waves have struck the islands. In the most destructive tsunami, 159 people died in Hilo on April 1, 1946, by killer waves generated by a powerful earthquake in the Aleutian Islands to the north.

Before the establishment of a tsunami watch, people had little advance warning of impending disaster except a rapid withdrawal of seawater from the shore. Residents of coastal areas frequently stricken by tsunamis have learned to heed this warning and head for higher ground. When a tsunami struck on the island of Maderia in the Azores during the 1755 Lisbon, Portugal, earthquake, large quantities of fish were stranded on shore as the sea suddenly retreated. Villagers, unaware of any danger, went out to collect this unexpected bounty, only to lose their lives when, without warning, a gigantic wave crashed down on top of them.

A few minutes after the sea retreats, a tremendous surge of water pounds the shore, extending hundreds of feet inland. Often a succession of surges occurs, each followed by a rapid retreat of water back to sea. On coasts and islands where the seafloor rises gradually or where barrier islands exist, much of the tsunami's energy is spent before it ever reaches shore. However, on volcanic islands surrounded by very deep water or where deep submarine trenches lie immediately outside harbors, an oncoming tsunami can build to tremendous heights.

Destructive tsunamis generated by large earthquakes can travel clear across the Pacific Ocean. The great 1960 Chilean earthquake elevated a California-sized chunk of land about 30 feet. It consequently created a 35-foot tsunami that struck Hilo, Hawaii, more than 5,000 miles away, causing more than $20 million in property damage and 61 deaths. The tsunami traveled an additional 5,000 miles to Japan and inflicted considerable destruction on the coastal villages of Honshu and Okinawa, leaving 180 people dead or missing. In the Philippines, 20 people were killed. Coastal areas of New Zealand were also damaged. For several days afterward, tidal gauges in Hilo could still detect the waves as they bounced around the Pacific Basin.

After discussing damaging earthquakes, the next chapter focuses on the destructiveness of volcanoes, including some of the most deadly eruptions in historic times, why they occur, and where they are likely to erupt in the future.

3

VOLCANIC ERUPTIONS

THE VENTING OF THE EARTH

This chapter examines volcanic activity and its dangers to civilization. Volcanoes are the second most destructive natural forces on Earth and constitute a major geologic hazard (Table 3). They are the most important land formers, producing numerous geologic structures from a variety of cones to huge lava flows. Volcanoes, whose eruptions are beneficial as well as hazardous, are the most spectacular of all of Earth's processes. They can be very damaging. Often a single eruption can wipe out entire towns and take the lives of thousands of people, as evident in the devastation wreaked on civilization down through the ages.

Two-thirds of all historic eruptions have caused fatalities. Since the Ice Age ended around 10,000 years ago, some 1,300 volcanoes are known to have erupted. During the past 400 years, more than 500 volcanoes have erupted, killing more than 200,000 people and causing billions of dollars in property damage. Since the year 1700, some two dozen volcanoes around the world have earned special recognition for killing more than 1,000 people each (Fig. 40). During the last 100 years, volcanoes have caused an average death toll of more than 800 people per year. In the 1980s alone, a decade of heightened volcanic activity, some 40,000 people lost their lives.

TABLE 3 MAJOR VOLCANIC ERUPTIONS

Date	Volcano	Area	Death Toll
1480 B.C.	Thera	Mediterranean	
A.D. 79	Vesuvius	Pompeii, Italy	16,000
1104	Hekla	Iceland	
1169	Etna	Sicily	15,000
1616	Mayon	Philippines	
1631	Vesuvius	Naples, Italy	4,000
1669	Etna	Sicily	20,000
1701	Fujiyama	Japan	
1759	Jorullo	Michoacan, Mexico	200
1772	Papandayan	Java, Indonesia	3,000
1776	Mayon	Philippines	2,000
1783	Laki	Iceland	10,000
1790	Kilauea	Hawaii	
1793	Unzen	Japan	50,000
1793	Tuxtla	Veracruz, Mexico	
1814	Mayon	Philippines	2,000
1815	Tambora	Sumbawa, Indonesia	92,000
1822	Galung Gung	Java, Indonesia	4,000
1835	Coseguina	Nicaragua	
1845	Nevado del Ruiz	Colombia	1,000
1845	Hekla	Iceland	
1850	Osorno	Chile	
1853	Niuafou	Samoa	70
1856	Pelée	Pierre, Martinique	
1857	St. Helens	Washington	
1873	Mauna Loa	Hawaii	
1877	Cotopaxi	Ecuador	1,000
1881	Kilauea	Hawaii	
1883	Krakatoa	Java, Indonesia	36,000
1886	Tongario	New Zealand	
1888	Bandai-san	Japan	460
1897	Mayon	Philippines	
1902	La Soufrière	St. Vincent, Martinique	15,000

TABLE 3 (CONTINUED)

Date	Volcano	Area	Death Toll
1902	Pelée	Pierre, Martinique	28,000
1902	Santa Maria	Guatemala	6,000
1903	Colima	Jalisco, Mexico	
1906	Vesuvius	Naples, Italy	
1910	Irazu	Costa Rica	
1911	Taal	Philippines	1,300
1912	Katmai	Alaska	
1912	Virunga	Belgian Congo	
1914	Lassen	California	
1914	Whakari	New Zealand	
1914	Sakurajima	Japan	
1917	San Salvador	El Salvador	
1919	Keluit	Java, Indonesia	5,500
1924	Kilauea	Hawaii	1
1926	Mauna Loa	Hawaii	
1927	Anak, Krakatoa	Java, Indonesia	
1928	Rokatinda	Dutch East Indies	
1929	Vesuvius	Naples, Italy	
1929	Calbuco	Chile	
1931	Merapi	Java, Indonesia	1,000
1932	Volcan del Fuego	Guatemala	
1932	Las Yeguas	Argentina	
1935	Kilauea	Hawaii	
1935	Mauna Loa	Hawaii	
1935	Coseguina	Nicaragua	
1938	Nyamlagira	Belgian, Congo	
1943	Parícutin	Michoacan, Mexico	
1944	Vesuvius	Naples, Italy	
1952	Binin Island	Japan	
1957	Capelinhos	Azores	
1963	Surtsey	Iceland	
1963	Agung	Bali, Indonesia	1,200

(continues)

TABLE 3 (CONTINUED)

Date	Volcano	Area	Death Toll
1969	Kilauea	Hawaii	
1973	Helgafell	Iceland	
1977	Nyiragongo	Zaire	70
1980	St. Helens	Washington	62
1983	El Chichon	Mexico	187
1983	Kilauea	Hawaii	
1985	Nevado del Ruiz	Armero, Colombia	22,000
1986	Lake Nios	Cameroon	20,000
1986	Augustine	Alaska	
1991	Pinatubo	Philippines	700
1991	Unzen	Japan	37
1994	Rabaul	Papua New Guinea	
1996	Ruapehu	New Zealand	
1997	Soufrière Hills	Montserrat	20
2000	Unzen	Japan	

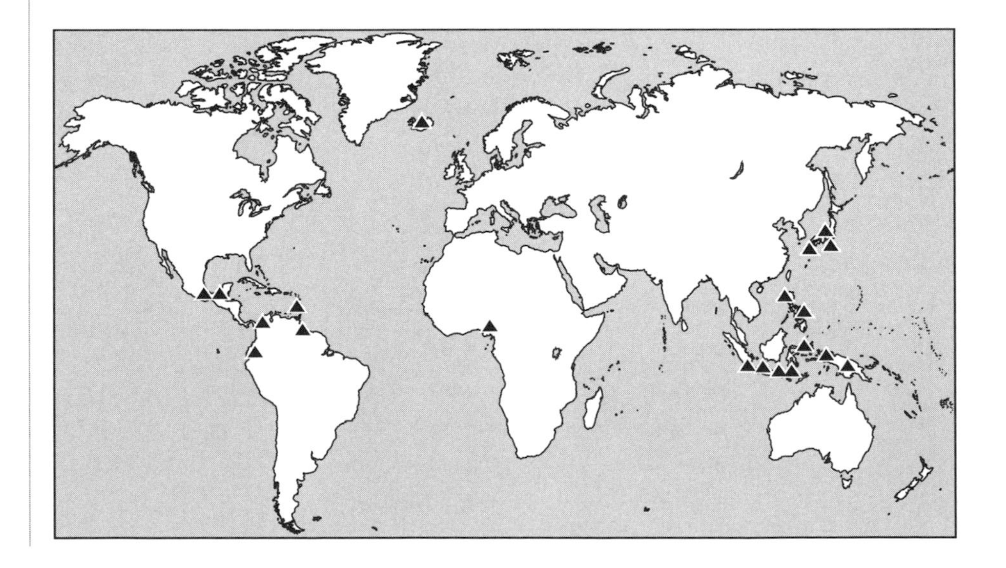

Figure 40 *Killer volcanoes responsible for the deaths of over 1,000 people each.*

The increase in volcano-related deaths is mainly due to swelling populations living near active volcanoes and not necessarily due to more eruptions. The majority of deaths occurs in developing countries that do not have the technology to provide advance warning. Presently, about 600 active volcanoes—meaning they have erupted in historic times—exist throughout the world. Many thousands more are dormant or extinct. However, even a volcano that has been dormant for up to a million years or more can suddenly awake without warning.

VIOLENT VOLCANOES

What could well be the most explosive volcanic eruption in recorded history took place in the Mediterranean Sea around 1625 B.C. on the island of Thera, 75 miles north of Crete. An outpost of the Minoan civilization lived and flourished there from 3000 B.C. until about 1480 B.C. The civilization collapsed so suddenly and violently that archaeologists have long been baffled. The eruption might have caused the demise of the Minoan civilization on Crete and surrounding islands.

A huge magma chamber beneath the island apparently flooded with seawater. Like a gigantic pressure cooker, the volcano blew its lid. The island then collapsed into the emptied magma chamber, forming a deep, gaping caldera that covered an area of about 30 square miles. The collapse of Thera created a colossal tsunami that battered the shores of the eastern Mediterranean, no doubt causing additional death and destruction. Although the initial toll on human life was probably only modest, the Minoan culture never completely recovered and rapidly declined.

The cataclysm was so violent it could have been seen in Egypt and heard as far away as Scandinavia. The Egyptians were almost certainly showered by a mysterious ash that blew in on the north wind. This immense ash cloud might have been responsible for the biblical Egyptian plagues. The weight of the thick ash might also account for the mysterious blow from the sky that destroyed Minoan palaces. Meanwhile, according to the book of Exodus, when Moses led the Israelites out of Egypt, "The Lord guided them by a pillar of cloud during the daytime and by a pillar of fire at night. The cloud and fire were never out of sight." Some historians believe the Greek poet Homer based some of his legends on the eruption, as did the Greek philosopher Plato, who ascribed his story about the lost island of Atlantis to the disaster.

Mount Vesuvius about 7 miles southeast of Naples, Italy, is perhaps the best-known volcano in the world. Presently, it is the only active one on the European mainland. Vesuvius lies within the rim of an earlier prehistoric volcano, called Mount Somma, and rises to a height of about 4,000 feet above sea level. The old, extinct volcano seemed not to pose any serious threat, even after several years of ground shuddering and wisps of steam rising from its crater. Then on August 24, A.D. 79, the seaward side of Vesuvius blew outward, sending searing hot ash mixed with steam and deadly gases toward the city of Pompeii. The volcanic debris buried people almost immediately. Caught in mid-stride, as many as 16,000 suffocated to death.

The Roman scholar Pliny the Elder lost his life to the eruption while studying the volcano a short distance away. His nephew, the Roman author Pliny the Younger, best describes the disaster. He wrote that a strange cloud shot up and outward from the mountain, the ground shook with violent earthquakes, and the waters of the Bay of Naples rose and fell several times and dashed upon the beaches. The mountaintop soon disappeared behind black, smoky steam that was pouring constantly from the crater with red flames and bright flashes of lightning darting through it. For eight days, the black cloud spread over towns around the mountain, and showers of hot pumice and globs of molten lava fell to earth.

Strangely, no lava flow was associated with this eruption as with succeeding outbursts. Instead, Pompeii was blanketed by 20 feet of ash, while the nearby town of Herculaneum was submerged in a sea of boiling mud. In 1740, the buried city of Pompeii was rediscovered by a farmer digging a water well. When archaeologists began excavating, they recovered various treasures and made casts of many inhabitants. The stark terror on the faces of the fossilized people was a gruesome reminder of the volcano's fury.

Another violent eruption of Vesuvius in 1631 destroyed several nearby villages and took 18,000 lives. The volcano struck again in 1661, taking the lives of 4,000 people in Naples. In 1906 and later in 1929, Vesuvius blew its top, devastating whole areas around it. The volcano erupted again during World War II on March 18, 1944 (Fig. 41). Vesuvius again put on a show of might, its most explosive eruption of the century. This time, the eruption threatened to interfere with the Allied invasion of Italy. The military advance was temporarily halted until several thousand inhabitants of Naples were evacuated to safety. Bombers of the 12th Army Air Force were caught on the ground near the base of the volcano, and lava flows caused extensive damage to the air base.

About 200 miles to the south, on the island of Sicily, lies Mount Etna, the largest and highest volcano in Europe, rising 10,902 feet above sea level. Records of Etna's eruptions kept by the early Greeks go back several centuries B.C. During its long history, the volcano's numerous outbursts have destroyed many towns and taken thousands of lives. The 1169 eruptions buried 15,000 people in the ruins of the nearby town of Catania. Exactly 500 years later, in 1669, 20,000 people perished. In 1928, the volcano wiped out the town of Mascati and almost entirely destroyed the village of Nunziata. Yet

Figure 41 *The 1944 eruption of Vesuvius Volcano, Naples, Italy.*
(Courtesy of USGS)

the townspeople refused to give up their vineyards and farms, for the volcanic soil is said to be among the richest in the world. Mount Etna produced another spectacular eruption in April 1992 when lava flows threatened the Sicilian town of Zafferana Etnea. People battled desperately against the fury of the volcano to turn the lava away from their homes by building berms to stem the flow.

Indonesian volcanoes (Fig. 42) are among the most explosive in the world and have produced more violent blasts in historic times than those of any other region. Mount Tambora on the island of Sumbawa began erupting on April 5, 1815, with a series of deep shocks that sounded like cannon fire and could be heard some 450 miles away. Dutch troops went to investigate what they thought were pirates attacking nearby military posts only to find a seething volcano. Then on April 11 and again on April 12, the volcano created one of the most explosive volcanic eruptions in the last 10,000 years, spewing some 25 cubic miles of debris into the atmosphere. The explosions were so powerful they could be heard in Sumatra, 1,000 miles to the west.

Only a few out of a total population of 12,000 survived on the island, and 45,000 or more additional lives were lost on adjacent islands. Heavy volcanic ash in sufficient quantities to darken the sky was carried as far as Java, 300 miles away. The eruptions caused climatic havoc, starvation, and disease around the world. The volcano cast a greater amount of volcanic dust into the air and obscured more sunlight than any eruption in the past 400 years. The lowered temperatures and killing frosts during the summertime resulted in "a year without summer." Crops in New England and western Europe failed, and famine threatened many parts of the northern hemisphere.

The volcanic island of Krakatoa, in the Sundra Strait between Java and Sumatra, produced the most destructive eruption in modern history. In May 1883, the dormant volcano began a series of violent explosions that were

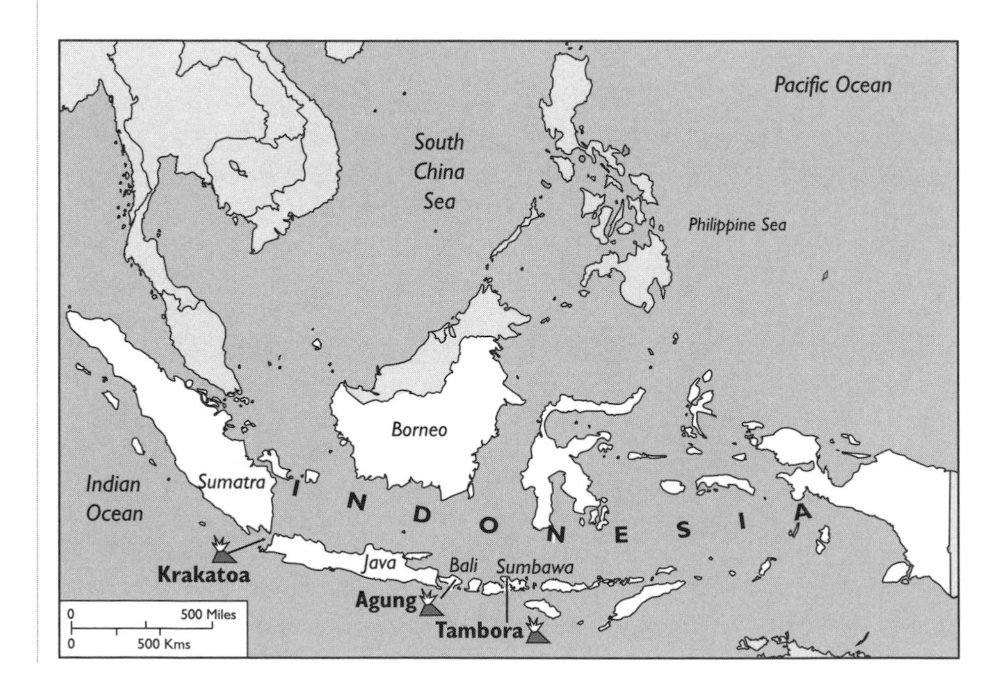

Figure 42 *Location of the great Indonesian volcanoes.*

heard 70 miles away and belched vast clouds of ash that obscured the sun. After dying down toward the end of May, large-scale eruptions began again in mid-June. This time, the eruptions were even more violent. Earth tremors were felt in many parts of Java and Sumatra. The main crater was blown away. In its place were as many as 10 new cones belching steam and pumice dust so high that ash fell onto villages 300 miles away. Then on August 27, the island was nearly totally decimated by a series of tremendous explosions that were heard as far away as Australia, Sri Lanka, and Madagascar.

The eruptions might have been powered by a rapid expansion of steam when seawater entered a breach in the magma chamber. Following the last convulsion, most of the island collapsed, creating a large undersea caldera more than 1,000 feet below sea level. The Krakatoa explosion is a classic example of how a large eruption can create a void in the magma reservoir, into which a roof several thousands of feet thick collapses, forming a huge gaping hole in the ocean floor.

The sound of the explosion carried as far as 3,000 miles away—it was called the loudest noise ever known. Barographs around the world recorded the atmospheric pressure wave as it circled the Earth at least three times. In the nearby coastal areas, the eruption produced towering tsunamis more than 100 feet high that swept 36,000 people to their deaths. For 100 miles around Krakatoa, the tsunamis ripped over low-lying areas, wiping out mainland towns and villages. The tsunamis were registered on tidal gauges as far away as the English Channel. Out of the ashes of Krakatoa arose a new volcanic island, which first erupted in June 1927, named Anak Krakatoa, or child of Krakatoa.

Mount Pelée on the island of Martinique in the West Indies was thought to be practically extinct, with the last relatively harmless eruption occurring in 1856. People long ago ceased to fear the volcano and looked upon it as a source of pride. Tourists even climbed the summit to enjoy the beautiful blue lake in its crater. On April 23, 1902, Pelée started to show signs of disturbance. A great column of smoke began to rise along with occasional showers of ash and cinders. Yet nothing indicated actual danger. On Monday, May 5, a torrent of steaming mud and lava burst through the crater and plunged into the valley below, destroying a sugar mill and killing several dozen workers.

On May 8, red flames were seen leaping from the mountain high into the sky, and a constant muffled roar occurred as enormous clouds of black smoke belched from the volcano. Then a tremendous explosion ripped out the seaward side of the 4,000-foot-high volcano. A solid sheet of flame rolled down the mountain and headed for the port city of St. Pierre. Clouds of hot ash and suffocating fumes called a nuée ardente swept through town and out to sea. The hot blast set fire to everything it touched (Fig. 43a, b), even ships in the harbor. Practically the entire population of 28,000 died within minutes. Nearly all the victims were found with their hands covering their mouths or

Figure 43a, b
Devastation at St. Pierre, Martinique, West Indies, from the May 8, 1902, eruption of Mount Pelée.
(Photo by I. C. Russell, courtesy USGS)

in some other agonizing posture, showing they had perished by suffocation. On the nearby island of St.Vincent, 15,000 people lost their lives to the eruption of La Soufríere that same year, during the 20th century's worst period of volcano-related fatalities.

Mount St. Helens, in Washington State, was an old, dormant volcano with a splendid, almost perfectly symmetrical cone. It lies in the Cascade Range, stretching from northern California to southern British Columbia, along with 15 other major active volcanoes. It has erupted at least 20 times in the past 4,500 years, and the last eruption was in 1980. The violent explosion on May 18, 1980, was the largest volcanic eruption in the continental United States in several centuries. The destruction was beyond imagination, with more than 200 square miles totally devastated by the volcano.

The eruption began when the north slope bulged out as much as 400 feet, creating avalanches on the mountain's upper flanks. Heat from the crater melted snow and formed numerous mudflows that scarred the slopes. From the weakened north flank, a lateral blast with hurricane force sent a superheated gas cloud racing down to the valley floor, devastating everything in its path for 18 miles. An immense landslide on the north side, one of the largest in historic times, plunged into Spirit Lake in the valley below. Lava flows, reaching speeds in excess of 70 miles per hour, periodically poured out of the crater and down the north flank into the valley.

The explosion blew off the top third of the mountain and lofted a cubic mile of debris into the atmosphere. The shock wave created by the explosion was felt as far away as Vancouver, British Columbia. Ash was carried northeastward by strong prevailing winds and fell as far as Montana, 600 miles away. Floods and mudflows from melting snow and rain swept through the valleys of the Toutle River. Mud-laden waters jammed with logs destroyed several bridges and clogged the channels of the Cowlitz and Columbia Rivers. At least 62 people lost their lives, and 200 were left homeless. Total damage was estimated at nearly $3 billion, including enough timber to build 80,000 houses when the nearby forest was flattened by the blast (Fig. 44).

El Chichon in Chiapas, southern Mexico, was thought to be extinct and had been dormant in historic times. Then on March 28, 1982, the volcano suddenly came alive (Fig. 45). El Chichon blasted a gigantic ash cloud high into the stratosphere, and several inches of ash covered nearby regions. The city of Palenque 75 miles east of the volcano bore the brunt of the ashfall, which measured more than 16 inches deep. Lava from the volcano buried nearby villages and killed 1,700 people, leaving 60,000 homeless. The last major eruption, occurring on April 4, sent a dense cloud of dust and gas into the stratosphere that was observed to travel in a narrow band completely around the world in three weeks. El Chichon produced notable effects on the climate as the ash cloud and volcanic gases significantly lowered global temperatures.

A similar type of eruption occurred at Mount Pinatubo, Philippines, on June 15, 1991, possibly the largest outburst of the 20th century. The eruption spewed about 20 million tons of weather-altering sulfur dioxide 22 miles into the atmosphere, about twice as much as El Chichon. Some 700 people died during the first three months of eruptions, and tens of thousands of families

Figure 44 *Trees destroyed by the lateral blast from the May 18, 1980, eruption of Mount St. Helens.*

(Photo by J. Hughes, courtesy USDA Forest Service)

lost their homes. Deep ash buried two important American military bases near the volcano, which had to be permanently abandoned. The eruption dropped global temperatures and was blamed for the strange weather that occurred worldwide in 1992, referred to as the "Pinatubo winter."

Figure 45 *Caldera formed by the March 28, 1982, massive eruption of the El Chichon Volcano, Chiapas, Mexico.*

(Courtesy USGS)

Nevado del Ruiz, Colombia, became the deadliest volcano in recent history. The mountain belongs to a chain of active volcanoes that runs through the western part of the country. The volcano has erupted at least a half-dozen times in the past 3,000 years, and an eruption in 1845 killed 1,000 people. The November 13, 1985, eruption created one of the world's largest volcanic mudflows called a lahar. When the Arenas crater erupted inside the 18,000-foot peak, it melted the mountain's ice cap. It sent floods and mudflows cascading 90 miles per hour down its sides into the nearby Lagunilla and Chinchina River valleys.

A 130-foot wall of mud and ash careened down the narrow canyon below and headed for the city of Armero 30 miles away. When the mudflow reached the town, 10-foot-high waves spread out and flowed rapidly through the streets. With the consistency of mixed concrete, the mudflow carried off everything in its path, including trees, cars, houses, and people. The mass of mud buried almost all the town, killing 23,000 people and leaving 60,000 others homeless. The mudflow also badly damaged 13 smaller towns where some 3,000 additional lives were lost. It was the second worst volcanic disaster of the 20th century. The eruption was similar to Mount St. Helens but with one very notable exception—this volcano was near a heavily populated area.

THE FIRE BELOW

Most of the world's active volcanoes concentrate in a few narrow belts (Fig. 46). The interaction of lithospheric plates on the Earth's surface is responsible for generating most of the world's volcanic activity. Subduction zones created by

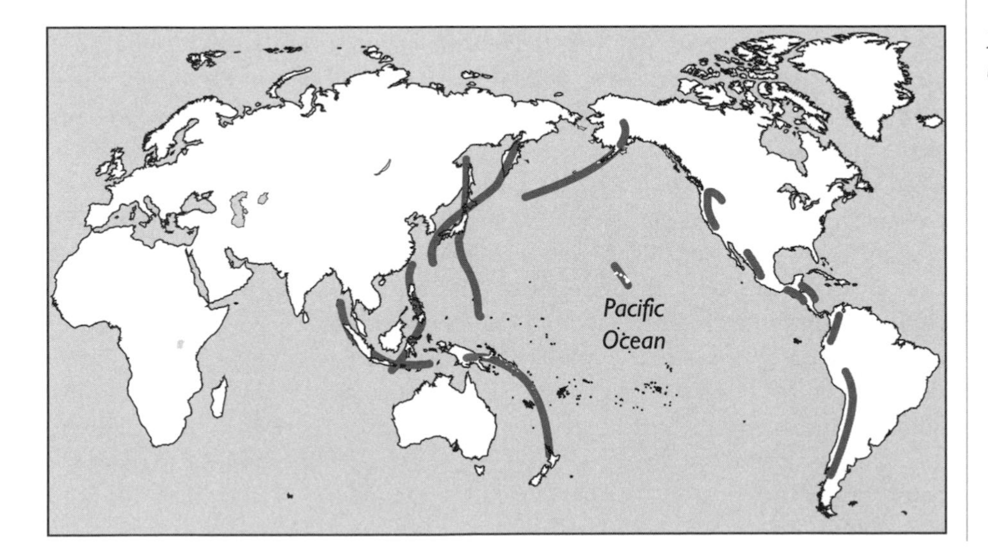

Figure 46 *Worldwide belts of active volcanoes.*

descending plates into the Earth's interior accumulate large quantities of sediment from adjacent continents and islands. The sediments are carried deep into the mantle, where they melt in pockets called diapirs. These rise toward the surface to form magma bodies that become the source for new igneous activity.

Some magma might also have originated from the partial melting of subducted oceanic crust, with heat supplied by the shearing action at the top of the descending plate. Convective motions in the wedge of asthenosphere caught between the descending oceanic plate and the continental plate force material upward, where it melts under lowered pressures.

The vast majority of mantle material that extrudes onto the surface is black basalt. Most of the 600 active volcanoes in the world are entirely or predominantly basaltic. The mantle material below spreading ridges, which create new oceanic crust, consists mostly of peridotite, which is rich in silicates of iron and magnesium. As the peridotite melts on its journey to the surface, a portion becomes highly fluid basalt. The magma that forms basalt originates in a zone of partial melting in the upper mantle more than 60 miles below the surface. The semimolten rock at this depth is less dense and therefore more buoyant than the surrounding mantle material and rises slowly toward the surface. As the magma ascends, the pressure decreases and more mantle material melts. Volatiles, such as dissolved water and gases, make the magma flow easily.

As the magma rises toward the surface, it replenishes shallow reservoirs or feeder pipes that are the immediate sources for volcanic activity. The magma chambers closest to the surface exist under spreading ridges where the oceanic crust is only 6 miles thick or less. Large magma chambers lie under fast-spreading ridges that create new lithosphere at a high rate, such as those in the Pacific. Conversely, narrow magma chambers lie under slow-spreading ridges, such as those in the Atlantic.

When the magma chamber swells with magma and begins to expand, it pushes the crest of the spreading ridge upward by the buoyant forces generated by the molten rock. The magma rises in narrow plumes that mushroom out along the spreading ridge, welling up as a passive response to plate divergence. Only the center of the plume is hot enough to rise all the way to the surface, however. If the entire plume were to erupt, it could build a massive volcano several miles high that would rival the tallest volcanoes on Mars (Fig. 47).

The composition of the magma indicates its source materials and the depth within the mantle from which they originated. Degrees of partial melting of mantle rocks, partial crystallization that enriches the melt with silica, and assimilation of a variety of crustal rocks in the mantle influence the composition of the magma. When the erupting magma rises toward the surface, it incorporates a variety of rock types along the way. This changes the magma's composition, which is the major controlling factor that determines the type of eruption.

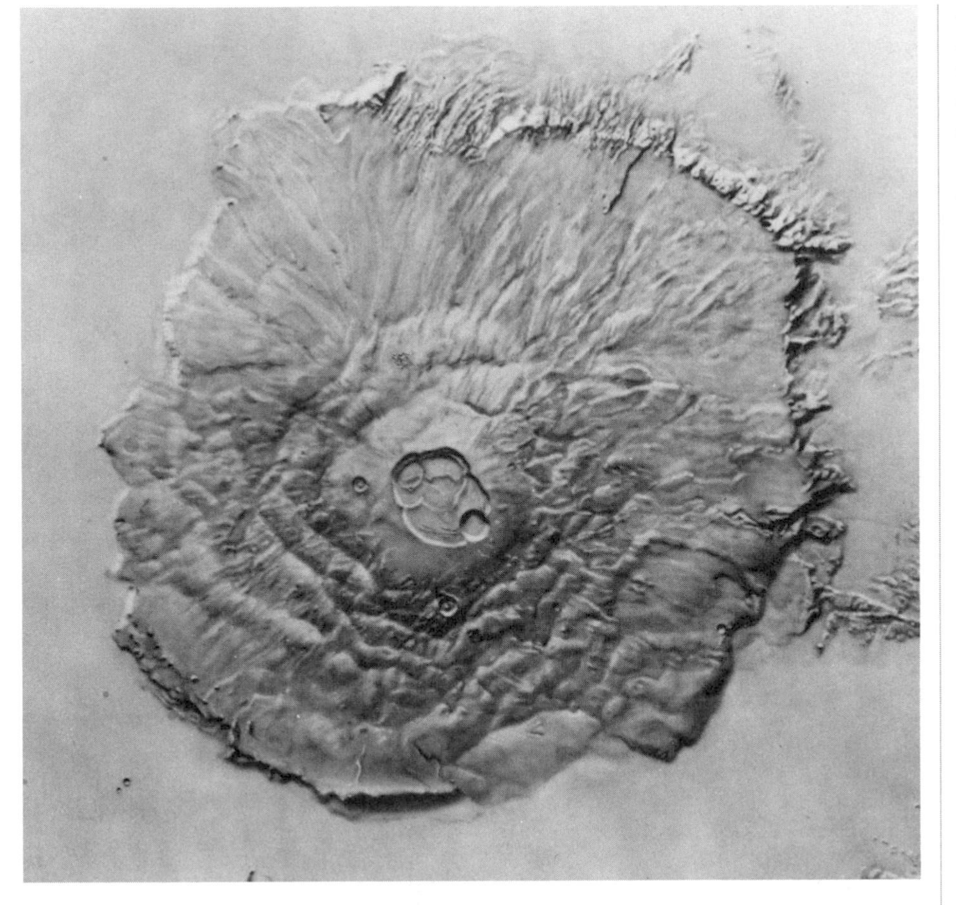

Figure 47 *Olympus Mons is the largest volcano on Mars, approximately 310 miles across and 17 miles high.*

(Photo by M. H. Carr, courtesy USGS)

The composition of the magma also determines its viscosity and whether it erupts mildly or explosively. If the magma is highly fluid and contains little dissolved gas when it reaches the surface, it produces basaltic lava. The eruption is then usually quite mild. The two types of lava from this type of eruption are aa, or blocky lava, and pahoehoe, or ropy lava. These are Hawaiian names and typical of Hawaiian volcanoes. However, if the magma rising toward the surface contains a large quantity of dissolve gases, it erupts in a highly explosive manner that can be quite destructive.

VOLCANIC ACTIVITY

The majority of volcanoes are associated with crustal movements at the margins of lithospheric plates. When one plate subducts under another, the lighter rock component melts and rises toward the surface in giant blobs of magma.

Figure 48 *The November 1968 eruption of Cerro Negro in west-central Nicaragua, which resembles a chain of volcanoes similar to the Cascade Range in the Pacific Northwest.*
(Courtesy USGS)

The molten rock subsequently feeds magma chambers lying below active volcanoes. Subduction zone volcanism builds volcanic chains on the continents and island arcs in the ocean. Subduction zone volcanoes, such as those in Indonesia in the western Pacific and along the western portions of Central and South America in the eastern Pacific, are among the most explosive volcanoes in the world (Fig. 48). Their violent nature is due to large amounts of volatiles, consisting of water and gases, in their magmas. As the magma rises toward the surface, the pressure drops, and volatiles escape explosively, shooting out of the volcano as though propelled by a gigantic cannon.

The Indonesian volcanoes Tambora and Krakatoa, produced by the subduction of the Australian plate down the Java Trench, are classic examples of subduction-zone volcanism. The 1991 highly explosive eruption of Pinatubo in the Philippines resulted from the subduction of the western Pacific plate down the Philippine Trench. Japan is a densely populated nation built around 86 active volcanoes. These include Mounts Unzen and Usu, which recently erupted due to the subduction of the Pacific plate down the Japan Trench. The Alaskan volcanoes Katmai and Augustine are noted for their massive ash eruptions due to the subduction of the Pacific plate down the Aleutian Trench.

The Cascade Range in the Pacific Northwest is a chain of powerful volcanoes associated with the Cascadia subduction zone, which is being overridden by the North American continent. The 1980 eruption of Mount St. Helens, whose blast leveled 200 square miles of national forest, is a good example of the explosive nature of these volcanoes. The eruption marked the beginning of a new episode of eruptive activity in the Cascades and other parts of the world. The decade of the 1980s witnessed the highest number of volcanic disasters and eruption-caused fatalities since 1902, when three eruptions in a six-month period killed more than 36,000 people.

The volcanic rock associated with subduction zones is a fine-grained, gray andesite, which contains abundant silica. This suggests a deep-seated source, possibly as much as 70 miles below the surface. The rock derives its named from the Andes Mountains, whose volcanoes are highly explosive due to subduction of the Nazca plate down the Peru-Chile Trench. As the magma rises toward the surface, the molten rock feeds volcanic magma chambers and buoys up the Andean mountain chain.

The second most common form of volcanism is rift volcanoes, which account for about 80 percent of all oceanic volcanism. Along spreading ridges, magma wells up from the upper mantle and spews out onto the ocean floor. The diverging lithospheric plates grow by the steady accretion of solidifying magma to their edges. More than 1 square mile of new oceanic crust comprising about 5 cubic miles of basalt is generated in this manner annually. At times, gigantic flows erupt onto the ocean floor, producing massive quantities of new basalt. Huge undersea fissure eruptions at spreading ridges on the ocean floor produce megaplumes of hot water. The ridge splits open and spills out hot water, while lava erupts in an act of catastrophic seafloor spreading.

Rift volcanoes are created by the divergence of lithospheric plates as the upper mantle is exposed to the surface. These volcanoes produce massive floods of basalt. The East African Rift Valley is a classic example. It extends from the shores of Mozambique to the Red Sea, where it splits to form the Afar Triangle in Ethiopia. Afar is perhaps one of the best examples of a triple junction created by the doming of the crust over a hot spot. The Red Sea and the Gulf of Aden represent two arms of a three-armed rift, with the third arm heading into Ethiopia. For the past 25 to 30 million years, the Afar Triangle has been stewing with volcanism and has alternated between sea and dry land.

The East African Rift is a complex system of tensional faults, indicating the continent is in the initial stages of rupture. During the process of rifting, large earthquakes rumble across the landscape as huge blocks of crust drop down along diverging faults. Much of the area has been uplifted thousands of feet by an expanding mass of molten magma lying just beneath the thinning crust. This heat source is responsible for the hot springs and volcanoes along the great rift valley, some of which are the largest and oldest in the world.

Iceland is a surface expression of the Mid-Atlantic Ridge. It is bisected by a volcanic rift responsible for the high degree of volcanism on the island. The rift produces a steep-sided, V-shaped valley flanked by several active volcanoes, making Iceland one of the most volcanically active places on Earth (Fig. 49). Often, volcanoes erupt under glaciers and melt the ice, causing massive floods to gush toward the sea. Geothermal energy produced by the volcanic activity provides more than 90 percent of the heat for buildings, without which Iceland would be an unbearably cold place to live.

Figure 49 *A lava flow that partially engulfed a building during the 1973 eruption on Heimaey, Iceland.*

(Courtesy USGS)

Another type of volcanism produces hot-spot volcanoes. These exist in the interiors of plates far from plate margins, where most volcanoes occur. They derive their magma from deep within the mantle, possibly to the very top of the core. The magma rises in giant mantle plumes that provide a steady flow of molten rock into magma chambers. The magma plumes rise through the mantle as separate giant bubbles of hot rock. When a plume passes the boundary between the lower and upper mantle, some 400 miles below the surface, the bulbous head separates from the tail and rises. This is often followed millions of years later by another similarly created plume, producing a one-two volcanic punch.

Perhaps the best example of hot-spot volcanism is the volcanoes that built the Hawaiian Islands. They appear to have been created by a single mantle plume beginning about 5 million years ago. The islands were assembled as though on a conveyor belt, with the Pacific plate traversing over the hot spot in a northwestward direction about 3 inches per year. Kilauea, whose Hawaiian name means "much spreading," has erupted almost continuously since 1983. Its lava flows have covered nearly 40 square miles and added about 300 acres of new land to Hawaii's coast. In effect, Kilauea has poured out more volcanic rock than the 1980 blast at Mount St. Helens.

Similar chains of volcanic islands in the Pacific trend in the same direction as the Hawaiian Islands, including the Line and Marshall-Gilbert Islands and the Austral and Tuamotu Seamounts. Seamounts form when magma flows from isolated volcanic structures that are strung out in chains across the interior of plates. Beneath the Pacific Ocean, more than 10,000 seamounts rise from the seafloor. However, only a few, such as the Hawaiian Islands and other islands crisscrossing the Pacific, manage to break the surface of the sea.

In the western Atlantic, the Bermuda Rise is oriented in a roughly northeast direction parallel to the continental margin off the eastern United States. It is nearly 1,000 miles long and rises some 3,000 feet above the surrounding seafloor, where the last of the volcanoes ceased erupting about 25 million years ago. A weak hot spot unable to burn a hole through the North American plate was apparently forced to take advantage of previous structures on the ocean floor to act as channels. This explains why the volcanoes trend nearly at right angles to the motion of the plate.

Off the west coast of Canada, the Bowie Seamount is the youngest in a line of submerged volcanoes running toward the northwest. It is fed by a mantle plume nearly 100 miles in diameter and more than 400 miles below the ocean floor. However, rather than lying directly beneath the seamount, as plumes are supposed to do, this one lies about 100 miles east of the volcano. The plume could have taken a tilted path upward, or the seamount could have moved with respect to the hot spot's position somehow.

On the continents, hot spots leave a distinct trail of volcanoes. A hot spot under North America is responsible for the volcanic activity in the Yellowstone Caldera, which covers an area of about 45 miles long by 25 miles wide. The hot spot can be traced across the Snake River Plain in southern Idaho. During the last 15 million years, the North American plate has traveled in a southwest direction over the hot spot, placing it under its temporary home at Yellowstone. In the last 2 million years, three major episodes of volcanic activity occurred in the region. They are counted among the greatest catastrophes of nature, and another major eruption is well overdue.

More than half of all hot spots exist under continents. They produce domes or swells in the crust up to 100 or more miles across, accounting for about 10 percent of the total land surface. Nearly all hot-spot volcanism occurs in regions of broad crustal uplift or swelling where magma lies near the surface. When a continental plate hovers over a number of hot spots, molten magma welling up from deep below creates domelike structures in the crust. The growing domes develop deep fissures through which magma rises to the surface. Africa has the largest concentration of hot spots, which are responsible for the unusual topography of the African continent, characterized by basins, swells, and uplifted highlands.

A single volcanic eruption can produce from a few cubic yards to as much as 5 cubic miles of volcanic material. This material is composed of lava and hot, solid products called pyroclastics (Fig. 50) along with large quantities

Figure 50 *A pyroclastic flow formed during the 1980 eruption of Mount St. Helens.*

(Courtesy USGS)

of water vapor and gases. Rift volcanoes account for about 15 percent of the world's active volcanoes and generate about 2.5 billion cubic yards of volcanic material per year, mainly submarine basalt flows. Some 20 eruptions of submarine rift volcanoes occur each year. Subduction zone volcanoes produce about 1 billion cubic yards of volcanic material per year, mostly pyroclastics. More than 80 percent of the subduction zone volcanoes, some 400 in all, exist in the Pacific. Hot-spot volcanoes produce about 500 million cubic yards of volcanic material per year, mostly basalt flows on the ocean floor and pyroclastics and lava flows on the continents.

Volcanoes have a variety of shapes and sizes, depending on the type of eruption. Cinder cones are relatively short with steep slopes. They form by explosive eruptions that deposit layer upon layer of large amounts of pumice and ash. One of the strangest volcanoes in recent times was formed in a cornfield on February 20, 1943, near the town of Parícutin, 200 miles west of Mexico City. Within just 10 weeks, the cinder cone attained a height of 1,100 feet and spread over hundreds of acres (Fig. 51).

If a volcano erupts only lava from a central vent or fissure, it forms a broad shield volcano, similar to Mauna Loa (Fig. 52) on the big island of Hawaii. Mauna Loa is the most voluminous volcanic mountain in the world. It has created a great sloping dome rising 13,675 feet above sea level and is made of some 25,000 cubic miles of lava. The basaltic lava spreads out, covering areas as large as 1,000 square miles. Kilauea, the youngest Hawaiian volcano, emerges from the flanks of Mauna Loa and, in time, might grow even larger than its parent. Mauna Kea to the north happens to be the world's tallest peak from base to tip. It rises more than 33,000 feet from the ocean floor and then climbs to a height of 13,800 feet above sea level.

Several dome-shaped volcanic features in northern California and Oregon, such as the Mono-Inyo Craters, are 3 or 4 miles wide and 1,500 to 2,000 feet high. Lava domes grow by expansion from within because the lava is too viscous or heavy to flow very far, causing it to pile up around the vent. Lava domes commonly occur in a piggyback fashion within the craters of large composite volcanoes. Good examples are California's Mono Domes and Lassen Peak.

Lava dome eruptions can be dangerous. They occur when sporadic eruptions extrude a pasty lava similar to toothpaste being squeezed from a tube. At Mount St. Helens, a lava dome began forming in late 1980. By the time volcanic activity ceased in 1986, it had grown to a height of about 900 feet. Lava domes pose a particular threat if they collapse and tumble down a volcano during an eruption, a common occurrence when they form on steep slopes. This happened in 1997 on the island of Montserrat in the West Indies during an eruption of the Soufrière Hills Volcano, which added only more misery onto an already embattled population.

Figure 51 *The June 28, 1943, eruption of Parícutin Volcano, Michoacan, Mexico.*

(Photo by W. F. Foshag, courtesy USGS)

A volcano that erupts a composite of cinder and lava is called a stratovolcano, which makes the tallest cones. Mount St. Helens is such a volcano. Highly explosive ash eruptions are followed by milder lava flows that reinforce the volcano's flanks, allowing it to grow to prodigious heights. Often these

Figure 52 *The Mauna Loa Volcano, Hawaii.*
(Courtesy USGS)

volcanoes end by decapitating themselves in a highly explosive eruption or by catastrophically collapsing into an empty magma chamber.

If such a cataclysm occurred at sea, it could generate a highly destructive tsunami. Volcanic eruptions that develop tsunamis are responsible for about a quarter of all deaths caused by tsunamis. The powerful waves can transmit the volcano's energy to areas outside the reach of the volcano itself. Large pyroclastic flows into the sea or landslides triggered by eruptions can also produce tsunamis. In 1792, during an earthquake after the eruption of Unzen, Japan, one side of the volcano collapsed into the bay. This created an enormous tsunami up to 180 feet high. It washed coastal cities out to sea, and as many as 15,000 people vanished without a trace.

GAS EXPLOSIONS

In northwest Cameroon of central Africa in a region of volcanic peaks and valleys covered by lush tropical vegetation, a deep crater lake known as Lake Nyos exploded on August 21, 1986. The gas eruption sent a deadly pall of toxic fumes, consisting of carbon dioxide, carbon monoxide, sulfur dioxide, hydrogen sulfide, and cyanide, spilling down the hillside. The gases spread out in a low, hanging blanket over a distance of more than 3 miles downwind from the lake. The hot, humid gases clung to people's clothing, which they frantically tried to discard. The gases immediately asphyxiated almost all villagers, and 20,000 people along with thousands of cattle and other animals lost their lives within a 10-square-mile area.

The disaster might have resulted from an earth tremor that cracked open the floor of the lake, releasing volcanic gases under great pressures. The gas discharge created a huge bubble that burst explosively through the surface of the lake and shot up to 500 feet into the air. This would explain why flattened plants were found more than 250 feet up on the slopes above the lake. The lake water was churned to a murky reddish brown from the stirred-up bottom sediments. In addition, the temperature of the water had risen 10 degrees Celsius. A similar eruption two years earlier at Lake Mamoum in the same range killed 37 people, suggesting that the disaster at Lake Nyos was no isolated occurrence.

On June 1, 1912, one of the 20th century's greatest eruptions produced a series of gigantic explosions that excavated a deep depression at the west base of Mount Katmai, Alaska. The depression was filled with a viscous, pancake-shaped lava that rose 800 feet in diameter and nearly 200 feet high. The lava covered an adjacent valley with a yellowish orange mass 12 miles long and 3 miles wide. Thousands of white fumaroles (volcanic steam vents) gushed from the lava flow, shooting hot water vapor as high as 1,000 feet into the air. Explorers discovering this wonder named the area the Valley of Ten Thousand Smokes.

Fumaroles are vents at the Earth's surface in volcanic regions that expel hot gases, often explosively. They exist on the surface of lava flows, in the calderas and craters of active volcanoes, and in areas where hot, intrusive magma bodies occur. The gas temperatures within the fumarole can reach 1,000 degrees Celsius. The primary requirement for the production of fumaroles and geysers is that a large, slowly cooling magma body lies near the surface to provide a continuous supply of heat. Yellowstone National Park is well known for its many fumaroles and geysers (Fig. 53).

The hot water and steam originate from juvenile water released directly from magma and other volatiles or from groundwater percolating downward near a magma body that heats the water by convection currents. Volatiles released from the magma body can also heat the groundwater from below. Generally, the bulk of the gases consists of steam and carbon dioxide, with smaller amounts of nitrogen, carbon monoxide, argon, hydrogen, and other gases. In a different type of fumarole, called a solfatara, from the Italian word meaning "sulfur earth," sulfur gases predominate.

Another result of the explosive release of trapped gases is blowouts. Hole in the Ground in the Cascade Range in Oregon is the site of a gigantic vol-

***Figure 53** Explosive burst of Seismic Geyser, Yellowstone National Park, Wyoming.*

(Photo by D. E. White, courtesy USGS)

canic gas explosion that created a huge crater. The crater is a near perfectly circular pit several thousand feet across with a rim raised several hundred feet above the surrounding terrain. For mysterious reasons, most of the crater lacks vegetation. Another volcanic structure called Ubehebe Crater is one of the most impressive sights in Death Valley, California. It exploded into existence about 1,000 years ago. When molten basalt came into contact with the shallow groundwater table, it flashed into steam, blowing out a huge hole in the ground.

Pockets of gas lie beneath the ocean floor, trapped under high pressure. As the pressure increases, the gases explode undersea and spread debris over wide areas, producing huge craters on the seabed. The gases rush to the surface in great masses of bubbles that burst when reaching the open air, resulting in a thick foamy froth on the ocean. In 1906, sailors in the Gulf of Mexico, southwest of the Mississippi Delta, witnessed such a gas blowout that sent mounds of bubbles to the surface.

Further exploration of the site revealed a large crater on the ocean floor below 7,000 feet of water. The elliptical crater measured 1,300 feet long, 900 feet wide, and 200 feet deep and sat on a small submarine hill. Downslope from the crater sat more than 2 million cubic yards of ejected debris. Apparently, the gases seeped upward along cracks in the seafloor and collected under an impermeable barrier. Eventually, as the pressure grew, the gas blew off its cover, forming a huge blowout crater. Similarly, the seafloor off Louisiana bears marks of numerous craters created by the eruption of buried salt deposits.

HAZARDOUS VOLCANOES

A number of volcanoes in the western United States are considered dangerous. Some, such as Mount St. Helens, might be awaking from long slumbers. Even 50,000 years of quietude are not long enough to silence the rumblings of volcanoes, some of which have awakened even after more than a million years of sleep. Forecasting future eruptions requires determining a volcano's past behavior by studying its rocks. The volcano can then be grouped along with others in a descending order of hazard.

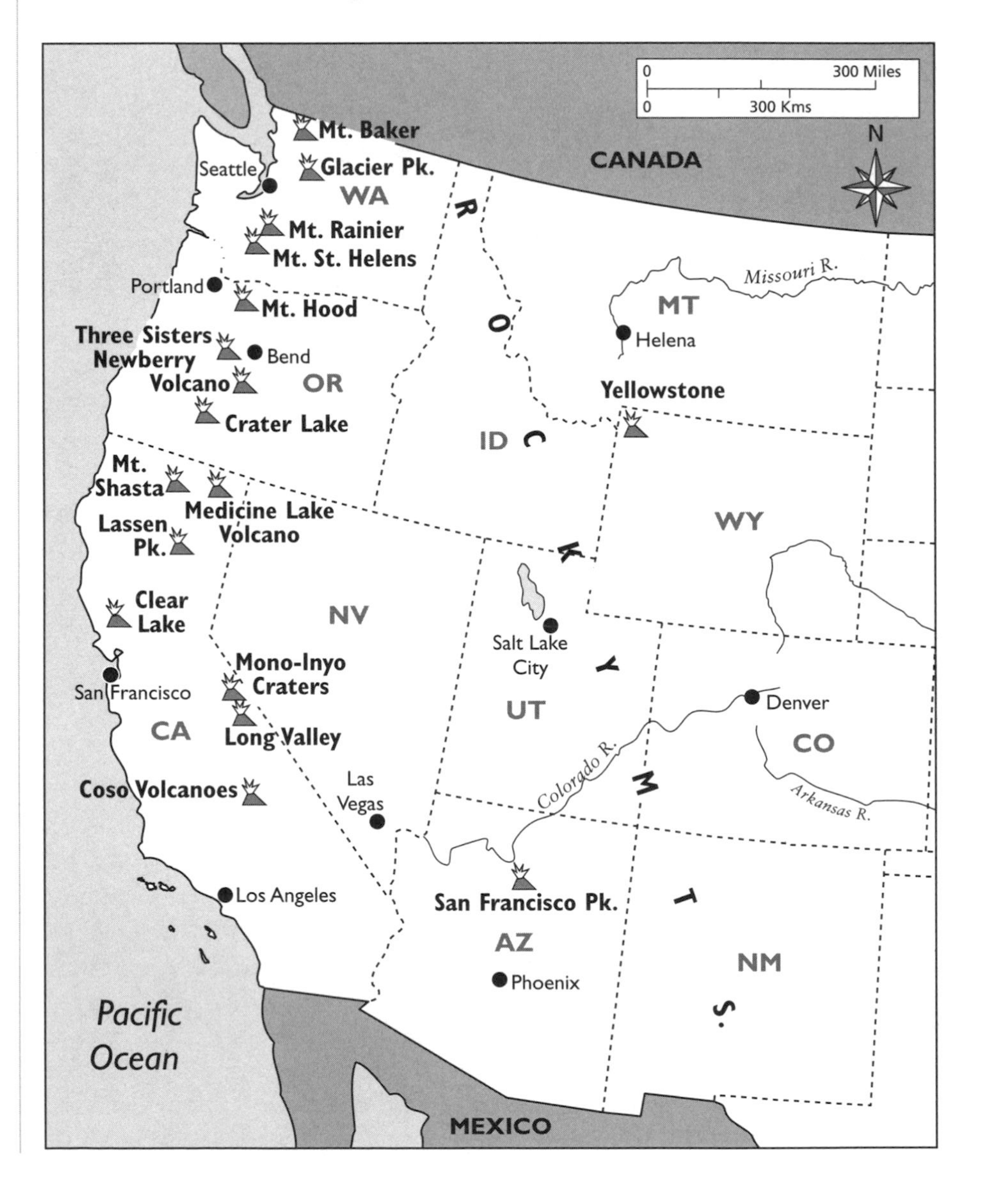

Figure 54 *Areas of potential future eruptions.*

More than 35 volcanoes in the United States, mostly in the Cascade Range, are likely to erupt sometime in the future (Fig. 54). The most hazardous are volcanoes that have erupted on average every 200 years, have erupted in the past 300 years, or both. These include, in descending hazardous order, Mount St. Helens, the Mono-Inyo Craters, Lassen Peak, and Mounts Shasta, Rainier, Baker, and Hood. Because of its location, looming above Seattle and Tacoma, Washington, Mount Rainier represents one of the greatest volcanic hazards in the United States. In the past, major sections of the volcano have collapsed, creating huge landslides and mudflows that swept through low-lying regions, presently home to 100,000 people.

The next most hazardous are volcanoes that erupt less frequently than every 1,000 years and last erupted more than 1,000 years ago. These include Three Sisters, Newberry Volcano, Medicine Lake Volcano, Crater Lake Volcano, Glacier Peak, and Mounts Adams, Jefferson, and McLoughlin. The third most hazardous are volcanoes that last erupted more than 10,000 years ago but still overlie large magma chambers. These include Yellowstone Caldera, Long Valley Caldera (Fig. 55), Clear Lake Volcanoes, Coso Volcanoes, San Francisco Peak, and Socorro, New Mexico.

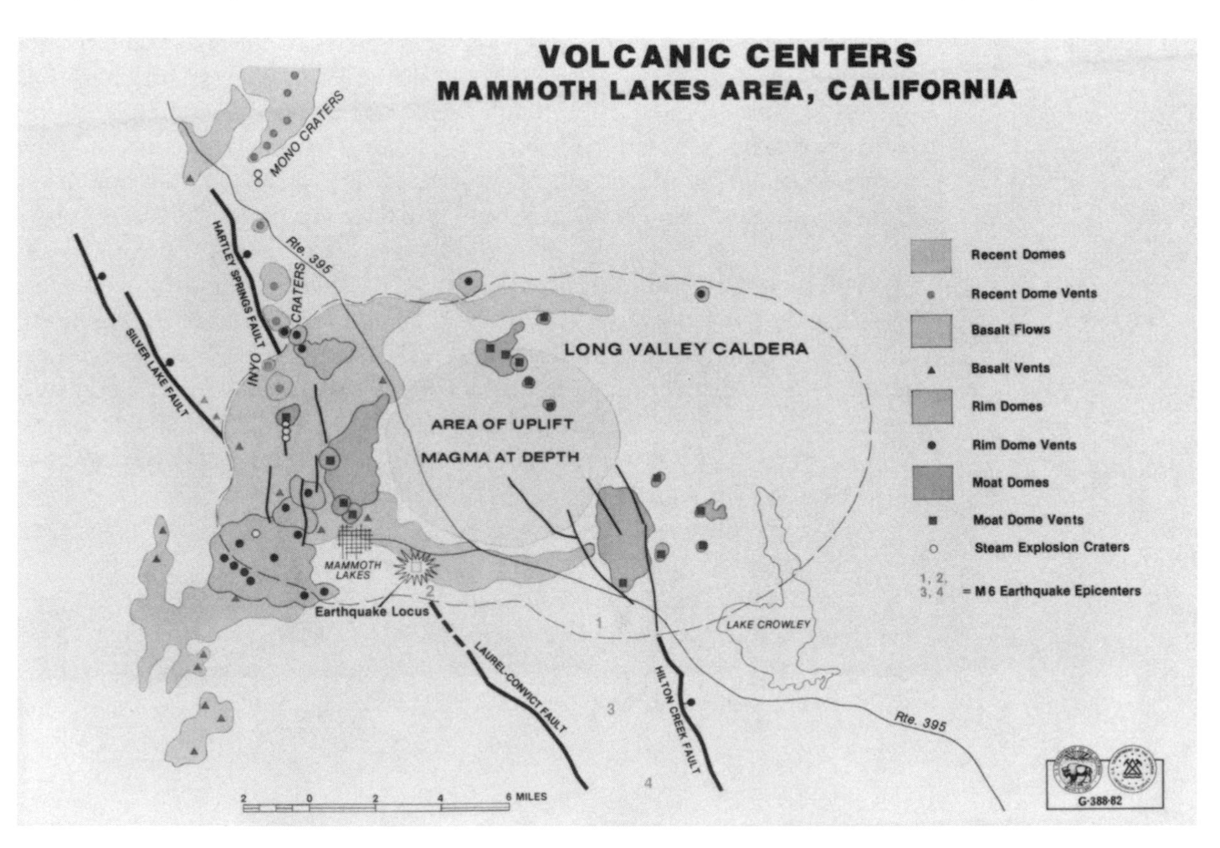

Figure 55 *A geologic map of the Long Valley Caldera, other volcanic centers, and major faults in the Mammoth Lakes area, Mono County, California.*

(Courtesy USGS)

Figure 56 *The May 22, 1915, eruption of Mount Lassen, Shasta County, California.*
(Photo by B. F. Loomis, courtesy USGS)

A map of geologically recent eruptions shows 75 centers of volcanic activity arrayed in broad bands. They extend from the Cascade Range in northern California, Oregon, and Washington eastward through Idaho to Yellowstone and along the border between California and Nevada. Another band extends from southeast Utah through Arizona and New Mexico. Because the pattern of activity 5 million years ago closely resembles the pattern since 10,000 years ago, all centers of activity have the potential for future eruptions. Additionally, new centers of activity might form within these bands at any time.

Historical records indicate that before the 1980 eruption of Mount St. Helens, only two other eruptions have occurred in the Cascade Range in the last century. A minor ash eruption took place at Mount Hood, Oregon, in 1906. Several spectacular eruptions of Lassen Peak, California, occurred between 1914 and 1917 (Fig. 56). Between 1832 and 1880, Mounts Baker, Rainier, St. Helens, and Hood erupted ash or lava. Periods between eruptions were 10 to 30 years for each volcano, and perhaps as many as three volcanoes erupted in the same year. However, none of these can compare with the eruptions of Mount St. Helens since its latest reawakening.

After discussing the how, why, and where of volcantic eruptions, along with related activities such as ash flows, mudflows, and floods, the next chapter investigates major geologic hazards arising from earth movements, including slides, flows, and erosion.

4

EARTH MOVEMENTS

THE FAILURE OF SLOPE MATERIALS

This chapter examines geologic hazards caused by ground failures. All earth movements such as landslides and related phenomena, including rockfalls, mudflows, earthflows, liquefaction, and subsidence, are naturally recurring events. However, they have become increasingly hazardous because people continue crowding onto susceptible lands.

Slopes are the most common and among the most unstable landforms. Under favorable conditions, the ground can give way even on the gentlest slopes, contributing to the sculpture of the landscape. Slopes are therefore only temporary features on the face of the Earth. Material on most slopes is constantly on the move. The rates vary from the imperceptible creep of soil and rock to catastrophic landslides and rockfalls that travel at tremendous speeds, often resulting in death and destruction.

LANDSLIDES

Landslides are rapid downslope movements of soil and rock materials under the influence of gravity (Fig. 57). They are triggered mostly by earthquakes

Figure 57 *The June 23, 1925, Gros Ventre landslide, Lincoln County, Wyoming.*

(Photo by W. C. Alden, courtesy USGS)

and severe weather. The main types of landslides are falls, topples, slides, spreads, and flows. Slides consisting of overburden alone are called debris slides, which are the most dangerous form of slope movement with respect to human life.

Slides involving bedrock with strong, resistant rocks overlying weaker beds are called rockslides and slumps. Slumps develop where unstable rocks are overlain by more sturdy rocks. Material slides downward in a curved plane, tilting up the resistant unit, while the weaker rock flows out and forms a heap. Slumps develop new cliffs just below preexisting ones, setting the stage for renewed slumping. Therefore, slumping is a continuous process. Usually many previous generations of slumps exist far in front of the present cliffs. This makes all cliffs inherently unstable and only temporary structures over geologic time.

Landslides occur in earth materials that fail along planes of weakness under shear (surface-to-surface contact) stress. They are initiated by an increase in shear stress and a reduction of shear strength in a rock formation, usually due to the addition of water to a slope. Water is almost always directly or indirectly involved with landslides. Water weathers rocks, which slowly reduces their shear strength near the surface. The weathering contributes to the instability of rocks, especially in areas with limestone, which dissolves in slightly acidic rainwater.

The slope geometry along with the composition, texture, and structure of the soil determines the formation strength. Changes in pore pressure and water content can weaken the friction between rock layers. The maximum natural inclination of a slope, called the angle of repose, is self-regulating by triggering slides that bring the slope back to its critical state when it becomes oversteepened. Therefore, the amount of sediment that accumulates balances the amount carried away by landslides.

Rock, soil, or snow particles pulled down a slope by gravity rub against each other and the ground as they fall. Each interaction causes particles to change direction and lose energy to friction. Usually, the shallower the slope angle, the lower the friction within the flow. Particles on the bottom of the slide in contact with the bed slow down, while the upper particles glide over them in a tumbling, chaotic mass. In this manner, the flow of material more closely resembles a dense gas of heavy, colliding particles rather than a true fluid.

Most landslides are not as spectacular as other violent forms of nature. However, they are more widespread, causing major economic losses and casualties in virtually every region of the world. In addition, landslides accompany other geologic hazards, including earthquakes, volcanic eruptions, and floods. The most severe landslides in the United States occur in mountainous regions, such as the Appalachian and Rocky Mountains and ranges along the Pacific Coast.

Coastal landslides occur when wave action undercuts a sea cliff, causing it to fall into the ocean. Sea cliff retreat results from marine and nonmarine agents, including wave attack, wind-driven salt spray, and mineral solution. The nonmarine agents responsible for sea cliff erosion include chemical and mechanical processes, surface drainage water, and rainfall. Mechanical erosion processes rely on cycles of freezing and thawing of water in crevasses, forcing apart fractures, which further weakens the rock.

Weathering breaks down rocks or causes the outer layers to shed by a process known as spalling. Animal trails that weaken soft rock and burrows that intersect cracks in the soil also erode sea cliffs. The sea cliff further erodes by surface water runoff and wind-driven rain. Excessive rainfall along the coast can lubricate sediments, enabling huge blocks to slide into the sea. Water running over the cliff edge and wind-driven rain produce the fluting often exposed on cliff faces.

Direct wave attack at the base of a sea cliff quarries out weak beds and undercuts the cliff. This causes the overlying unsupported material to collapse onto the beach. Waves also work along joint or fault planes to loosen blocks of rock or soil. In addition, winds carrying salt spray from breaking waves drive it against the sea cliff. Porous sedimentary rocks absorb the salty water, which evaporates and forms salt crystals whose growth weakens rocks. The surface of the cliff slowly flakes off and falls to the beach below, where the material landing at the base of the sea cliff piles up into a heap of talus.

The direct costs from landslide damage to highways, buildings, and other facilities as well as indirect costs resulting from the loss of productivity can run

Figure 58 *Sliding of unstable earth materials undermine the foundations of houses in the Pacific Palisades area of Southern California.* (Photo by J. T. McGill, courtesy USGS)

into the billions of dollars annually. Single large slides in populated areas can cost tens of millions of dollars. Fortunately, landslides in the United States have not resulted in a major loss of life such as those in other parts of the world because most catastrophic slope failures in this country generally occur in sparsely populated areas. One example is a slide that buried a highway and railroad tracks in the Wasatch Range of Utah during heavy spring rains in 1983. The slide created a dam that formed a lake, which threatened residents below with flooding, forcing the evacuation of 500 people.

The exception occurs in California, whose residents are well aware of landslides in their state (Fig. 58). Thousands of landslides causing considerable property damage have occurred in the Los Angeles basin alone. Often, repeated heavy rains and floods devastate the hillsides of Los Angeles, setting off landslides that destroy or seriously damage homes perched on unstable ground. In response to this year-round destruction, officials began dealing with nonearthquake geologic hazards in the mountainous and hilly seaside regions. They passed landslide control legislation that requires new building sites to be inspected by an accredited geologist.

The majority of landslides are triggered by earthquakes. The size of the area affected by these landslides depends on the magnitude of the temblor, the topography and geology of the ground near the fault, and the amplitude and duration of the ground motion. In past earthquakes, landslides have been numerous in some areas having intensities of ground shaking as low as

3 to 5 magnitude. During the 1959 Hebgen Lake, Montana, earthquake, which killed 26 people, a single large slide, moving from north to south, gouged out a huge scar in the mountainside (Fig. 59). Debris traveled uphill on the south side of the valley and dammed the Madison River, creating a large lake.

Earthquake-induced landslides often result in wide area destruction. In the 1964 Good Friday Alaskan earthquake, landslides and ground subsidence caused the greatest damage to buildings and other structures. The ground beneath Valdez and Seward literally gave way. Both waterfronts floated toward the sea, taking with them the lives of 31 people. In Anchorage, landslides caused $50 million in damage, and houses were destroyed when 200 acres were carried toward the ocean. The destruction was so complete that the area was bulldozed over and made into a monument, appropriately named "Earthquake Park."

Figure 59 *The August 1959 Madison Canyon slide, Madison County, Montana.*

(Photo by J. R. Stacy, courtesy USGS)

The 1971 San Fernando, California, earthquake produced nearly a thousand slides distributed over a 100-square-mile area of remote and hilly mountainous terrain. The 1976 Guatemala City earthquake triggered some 10,000 landslides throughout an area of 6,000 square miles. On March 5, 1987, during the rainy season in Ecuador, an earthquake shook loose fierce mudslides that buried villages in the rugged hilly region, killing more than 1,000 people.

Landslides can also be triggered by the removal of lateral support by erosion from streams, glaciers, waves, or currents. They are initiated by previous slope failures and human activities, such as excavation and other forms of construction. The ground also gives way under excess loading by the weight of rain, hail, or snow. In addition, the weight of buildings and other structures tends to overload a slope, causing it to fail.

Other common triggering mechanisms for landslides are dynamite explosions that break the bond holding the slope together, overloading the slope so it can no longer support its new weight, undercutting at the base of the slope, and oversaturation with water from rain or melting snow. Water adds to the weight of the slope and decreases the internal cohesion of the overburden. The effect of water as a lubricant is very limited, however. Its main effect is the loss of cohesion when the spaces between soil grains are filled with water.

In volcanic mountainous regions, seismic activity and uplift associated with an eruption cause landslides in thick deposits of unconsolidated pyroclastic material on a volcano's flanks. The distribution of landslides in volcanic terrain is determined by the seismic intensity, topographic amplification of the ground motion, rock type, slope steepness, and fractures and other weaknesses in the rock. Heavy sustained rainfall over a wide area also triggers landslides and mudflows.

Even dormant volcanoes can be unstable, causing entire slopes to collapse. Volcanic rocks that were once hard tend to break down and soften with time, causing slopes to collapse under their own weight. Volcanoes themselves also put tremendous pressure onto the sediments below, causing them to seep out from beneath the volcano on one side. This creates uneven stresses on the volcano, which can lead to fractures and ultimately a massive collapse and landslides. Many volcanoes capable of collapsing exist near populated areas, such as Seattle and Mexico City.

When Mount St. Helens erupted in 1980, a wall of earth slid down the mountain in what was called the greatest landslide in modern history (Fig. 60). It filled the valley below with debris measuring 5 miles by 4 miles. One arm of the gigantic mass plowed through Spirit Lake at the base of the volcano and burst into the valley beyond, devastating everything in its path for 18 miles. Massive mudflows scoured the slopes of the volcano and jammed the

Figure 60 *Area affected by massive landslides and mudflows from the May 18, 1980, Mount St. Helens eruption, showing Spirit Lake in the foreground.*

(Photo by Jim Hughes, courtesy USDA Forest Service)

Cowlitz and Columbia Rivers to the Pacific Ocean with debris and timber blown down by the tremendous blast.

The longest landslide ever reported tumbled off Mexico's Nevado de Colima Volcano more than 18,000 years ago. The Colima slide sped 75 miles to the Pacific Coast and then some distance into the ocean. The western rim of Mount Etna's Valle del Bove caldera resulted from a collapse of the mountain's eastern slope centuries ago. Ground motion has destabilized the caldera, which could lead to rockslides, mudslides, or even a full-scale eruption.

Thousands of years ago, Mount St. Augustine (Fig. 61) built an obscure island out of lava and ash in the Cook Inlet 175 miles southwest of Anchorage, Alaska. Every 150 to 200 years, large parts of the volcano have collapsed and fallen into the sea, generating large tsunamis. Massive landslides have ripped out the volcano's flanks 10 or more times during the past 2,000 years. The most recent slide occurred during the October 6, 1883, eruption when debris on the flanks of the volcano crashed into the Cook Inlet. This sent a 30-foot tsunami to Port Graham 54 miles away that destroyed boats and flooded houses. Subsequent eruptions have filled the gap left by the last landslide, making the volcano increasingly unstable and setting the stage for another collapse. If a landslide does occur, it would barrel down the north side of the volcano and plunge into the sea. This would send a tsunami in the direction of cities and of oil platforms residing in the inlet.

On March 27, 1986, Mount St. Augustine awoke after 10 years of slumber. The initial eruption shot ash 9 miles into the atmosphere, while the volcano continued steadily pumping ash and gases. The ash cloud spread as far as 600 miles to the north, and blinding ash in the town of Kenai kept motorists off the streets. Ash was thick enough at the coastal town of Homer, 70 miles east of the volcano, for the streetlights to come on. Tsunami warnings were given in case a violent eruption sent debris on

Figure 61 *Mount St. Augustine, Kamishak District, Cook Inlet, Alaska.*

(Photo by C. W. Purington, courtesy USGS)

the flanks of the volcano crashing into the inlet as they have often done in the past.

Another type of downslope movement is volcanic pyroclastic flows, which consist of masses of hot, dry rock debris that move rapidly downslope like a fluid. Their mobility is due to hot air and other gases mixed with the debris called a nuée ardente, French for "glowing cloud." They often form when large masses of hot rock fragments suddenly erupt onto a volcano's flanks. Pyroclastic flows travel at speeds up to 100 miles per hour and tend to follow valley floors, burying them with thick deposits. Because of their great mobility, pyroclastic flows can affect areas 15 miles or more away from a volcano. The swiftly moving hot rock debris buries and incinerates everything in its path. Clouds of dust and gas can blanket adjacent areas downwind, burning and asphyxiating people and animals.

Landslides are often called avalanches, but this term is generally reserved for snowslides. Avalanches usually begin with a mass of fresh, powdery snow resting on a steep bank of older snowpack and triggered by earthquakes, loud noises, or skiers. A spectacular example of an avalanche took place in the Andes Mountains of Peru on May 31, 1970. A 7.7 magnitude earthquake triggered a sliding mass of glacial ice and rock, 3,000 feet wide and about a mile long, that rushed rapidly downslope with a deafening roar, accompanied by a strong, turbulent air blast. Frictional heat partially melted the ice, making the slopes even more slippery. The avalanche traveled nearly 10 miles to the town of Yungay in four minutes or less, burying it under thousands of tons of rubble.

Thousands of boulders, weighing up to several tons, were hurled more than 2,000 feet across the valley (Fig. 62). Their presence indicated the slide must have reached a velocity of nearly 250 miles per hour. The volume and velocity of this enormous plunging mass enabled it to ride over obstacles such as a 600-foot ridge between the valley and Yungay. The slide shot across the valley and 175 feet up the opposite bank, where it partly destroyed another village. Flash flooding from broken mountain lake basins and from the avalanche-swollen waters of the Río Santa created a wave as high as 45 feet, which caused flooding that only exacerbated the disaster. When the slide ended, 18,000 people had lost their lives to the earthquake's death and destruction (Fig. 63).

On January 16, 1995, a blizzard in the foothills of the Himalayas in Kashmir, northern India, stranded hundreds of people who abandoned their cars and busses on a one-lane highway to take shelter inside a 1.5-mile-long tunnel. Without warning, an avalanche struck, burying everything in the area. Some people managed to escape the tunnel before thousands of tons of snow completely closed it off. Several days later, bulldozers and villagers armed with shovels dug through the wall of snow, only to find, to their horror, the tunnel filled with frozen bodies.

Figure 62 *A large boulder transported by the May 31, 1970, avalanche in the Andes Mountains, Peru.*

(Courtesy USGS)

Figure 63 Destruction of adobe houses in Huaraz, Peru, from an avalanche crated by the May 31, 1970, Peruvian earthquake.

(Courtesy USGS)

ROCKSLIDES

Rockslides (Fig. 64) are generally large and destructive, often involving millions of tons of rock. They are created by a mass of bedrock that breaks into many fragments during the fall. The material behaves as a fluid, spreading out onto the floor below. The slide might have enough energy to flow some distance uphill on the opposite side of a valley. Rockslides are prone to develop when planes of weakness, such as bedding planes or jointing, are parallel to a slope. This is especially true if the slope has been undercut by a river, glacier, or construction work.

During one of the most devastating rockslides, the town of Elm, Switzerland, was wiped off the map on September 11, 1881, when a nearby mountainside suddenly collapsed, transforming a solid cliff into a river of rock. The collapsing cliffside plummeted 2,000 feet and sped through the valley below for a distance of nearly 1.5 miles. As the gigantic mass of rock debris rapidly roared down the valley, it entombed 116 people beneath a thick blanket of broken slate before grinding to a halt.

Soon after the slide, Swiss geologist Albert Heim visited the site. He discovered that such large slides, called long-runout landslides, travel for long, horizontal distances because they encounter little base-level friction to slow them down. Although large slides occur infrequently, they can potentially cause great destruction because they travel so far, obliterating towns that seemed to be a safe distance from precarious mountain slopes. Most small landslides move horizontally less than twice the distance they fall. However, large landslides can travel as much as 10 times as far as they fall.

Figure 64 *A rockslide that blocked Highway 1, South of Big Sur, California, during severe winter storms in 1982–1983.*

(Photo by G. F. Wieczorek, courtesy USGS)

The Blackhawk slide at the southern edge of the Mojave Desert, about 85 miles east of Los Angeles, California, is one of the more spectacular examples of a long-runout landslide. It occurred roughly 17,000 years ago when a large chunk of Blackhawk Mountain collapsed. A mass of marble rock fell about a mile and spread horizontally over the essentially flat desert land for nearly 6 miles, attaining speeds of 75 miles per hour. The slide flowed like a fluid, spilling across the desert in a dark, fan-shaped sheet. Apparently, the slide spread so far because it rode on top of a layer of air, which acted as a lubricant. An alternative explanation suggests that bottom particles bounced around so energetically they supported the bulk of the landslide.

One of the most destructive rockslides took place on October 9, 1963, at Monte Toc in the Italian Alps, named by the local residents as "the mountain that walks." Despite efforts to stabilize the slopes, the mountain not only walked, it galloped. The mountain anchored one side of the newly constructed Vaiont Dam, whose reservoir was only half filled with water. A torrent of water, mud, and rock plunged into the narrow gorge, shot across the wide bed of the Piave River, and ran up the mountain slope on the opposite side. The slide completely demolished the town of Longarone, killing 2,000 of its residents.

This was a nonearthquake-produced landslide that was called history's greatest dam disaster. Yet when the violence was over, the dam was still intact. Some 600 million tons of debris from the mountainside slid instantaneously into the reservoir. The water was forced 800 feet above its previous level, rising in one great wave 300 feet above the dam before dropping into the gorge below. The water quickly picked up speed when constricted by the narrowness of the gorge, snatching tons of mud and rock as it raced on its destructive journey downstream.

During the October 17, 1989, Loma Prieta, California, earthquake, rockslides rumbled out of the Santa Cruz Mountains and buried highways (Fig. 65). The earthquake broke along a segment of the San Andreas Fault that runs through the mountain range, which shifted dramatically during the temblor. The earthquake raised the southwest side of the fault more than 3 feet, which substantially contributed to the continued growth of the Santa Cruz Mountains.

Another rockslide southeast of Glacier Point in Yosemite National Park, California, on July 10, 1996, sent 160,000 tons of granite that broke off a cliff plunging a third of a mile at more than 160 miles per hour. The slide caused a hurricane-like air blast that leveled thousands of trees, some with their bark completely stripped off. The air blast represents a poorly understood collateral hazard of rockfalls similar to dropping a book parallel to the ground, which forces the air out from beneath it. As a result, geologists might have to reassess hazard zones marked on maps at Yosemite and other mountainous national parks to take into account the danger from air blasts.

Figure 65 *A large rockslide near the summit of the Santa Cruz Mountains from the October 17, 1989, Loma Prieta, California, earthquake.*

(Photo by G. Plafker, courtesy USGS)

If material drops off a nearly vertical mountain face at the velocity of free fall, it results in a rockfall or soilfall, depending on its composition. Rockfalls range in size from individual blocks plunging down a mountain slope to the failure of huge masses of rock weighing hundreds of thousands of tons falling nearly vertically down a mountain face. Individual blocks commonly come to rest at the base of a cliff in a loose pile of angular blocks, often forming a talus cone (Fig. 66).

Immensely destructive waves are set in motion if large blocks of rock drop into a standing body of water, such as a lake or fjord. A 1958 earthquake in Alaska triggered an enormous rockslide that fell into Lituya Bay, generating a gigantic wave that surged 1,700 feet up the mountainside. Trees toppled over like matchsticks when massive quantities of seawater inundated the shores (Fig. 67). Coastal landslides of large magnitude can also generate destructive tsunamis. This hazard is particularly feared in Norway, where small deltas along fjords might provide the only available flat land at sea level. Waves generated by rockfalls can range from 20 to 300 feet high and cause considerable damage as they burst through local villages.

The most impressive rockfall ever recorded occurred at Gohna, India, in 1893. A huge mass of rock, loosened by driving monsoon rains, dropped 4,000 feet into a narrow Himalayan mountain gorge, forming a gigantic dam 3,000 feet across and 900 feet high that extended 11,000 feet upstream and downstream. The huge pile of broken rock, measuring about 5 billion cubic yards, impounded a large lake 770 feet deep. Two years after the fall, the dam burst,

Figure 66 *Large talus cones in the Stinking Water Canyon, Park County, Wyoming.*

(Photo by T. A. Jaggar Jr., courtesy USGS)

Figure 67 *The trimline shown in light areas, where trees were destroyed by an enormous rockslide into Lituya Bay, Alaska, in 1958.*

(Courtesy USGS)

causing a world-record flood. Some 10 billion cubic feet of water discharged within hours, producing floodwaters that crested 240 feet high.

The largest rockfall in the Rocky Mountain region in several thousand years occurred in Alberta, Canada, on April 29, 1903. It might have been triggered by coal mining below the base of Turtle Mountain, which weakened the mountain. A mass of strongly jointed limestone blocks on the crest of the mountain broke loose and plunged down the deep escarpment. About 90 million tons of material tumbled down the mountainside and swept through the town of Frank in one tremendous wave, killing 70 people. The rockfall then swept up the opposite slope 400 feet above the valley floor.

SOIL SLIDES

Earthquakes can cause soil slides in weakly cemented, fine-grained materials that form steep, stable slopes. Many slopes failed during the New Madrid, Missouri, earthquakes of 1811–1812, when massive landslides slid down bluffs and low hills. The 1920 Kansu, China, earthquake produced soil flow failures that killed an estimated 180,000 people. As the tremor rumbled through the region, immense slides rushed out of the hills, burying entire villages and damming streams that flooded valleys.

During heavy rainstorms, soil on a steep hillside can suddenly change into a wave of sediment, sweeping downward at speeds of more than 30 miles per hour. Precipitation frees dirt and rocks by increasing the water pressure in pores within the soil. As the water table rises and pore pressure increases, friction holding the top layer of soil to the hillside diminishes until the pull of gravity overcomes it. Immediately before the soil begins to slide, the pore pressure drops and the soil starts to expand.

Creep (Fig. 68) is a slow downslope movement of soil, often recognized by downhill-tilted poles, fence posts, and trees. It is a more rapid movement of near-surface soil material than the sediment below and occurs particularly rapidly where frost action is prominent. After a freeze-thaw sequence, material moves downslope due to the expansion and contraction of the ground. In frozen terrain, permafrost is impermeable to water, causing the overlying saturated soil to slide downhill under the force of gravity. Soil that has been weakened by frost action is most susceptible to this process. In temperate regions, trees are unable to root themselves, and only grasses and shrubs can grow on the slope. If creep is especially slow, tree trunks are bent. After the trees become tilted, new growth attempts to straighten them. If the creep is continuous, trees lean downhill in their lower parts and become progressively straighter higher up.

An earthflow (Fig. 69) is a more visible form of movement caused by raised water content of the overburden, which increases the weight and reduces the sta-

Figure 68 *Railroad track bent by creep in the Nome River Valley, Alaska.*

(Photo by R. S. Sigafoos, courtesy USGS)

bility of the slope by lowering resistance to shear. Earthflows are characterized by grass-covered, soil-blanketed hills. Although they are generally minor features, some can be considerably large, covering several acres. Earthflows usually have a spoon-shaped sliding surface upon which a tongue of overburden breaks away and flows for a short distance, forming a curved scarp at the breakaway point.

Figure 69 *An earthflow on slopes west of Prosperity, Washington County, Pennsylvania.*

(Photo by J. S. Pomeroy, courtesy USGS)

Expansive soils are sediments that swell or shrink due to changes in moisture content. Expansive soils are abundant in geologic formations in the Rocky Mountain region, the Basin and Range Province, the Great Plains, the Gulf Coastal Plain, the lower Mississippi River Valley, and the Pacific Coast. The parent materials for expansive soils are derived from volcanic and sedimentary rocks that decompose to form expansive clay minerals such as montmorillonite and bentonite. These materials are often used as drilling muds because of their ability to absorb large quantities of water. Unfortunately, this characteristic also causes them to form highly unstable slopes. Swelling soils produce the greatest annual losses of all geologic hazards. Damages to buildings and other structures built on expansive soils cost the United States several billion dollars annually.

MUDFLOWS

Mudflows (Fig. 70) are among the most impressive features of desert regions. Heavy runoff in the bordering mountains forms rapidly moving sheets of water that pick up huge quantities of loose material. The floodwaters flow into a stream, where all the muddy material suddenly concentrates in the channel. The dry streambed rapidly transforms into a flash flood that moves swiftly downhill, often with a steep, wall-like front. The behavior of mudflows is similar to that of a viscous fluid, often carrying a tumbling mass of rocks and large boulders.

Figure 70 *The Slumgullion mudflow into Lake San Cristobal in September 22, 1905, Hinsdele County, Colorado.*

(Photo by W. Cross, courtesy USGS)

Mudflows can cause considerable damage as they flow out of mountain ranges. At the base of the range, velocity drops. The loss of water by percolation into the ground thickens the mudflow until it comes to a complete halt. Mudflows can carry large blocks and boulders the size of automobiles onto the floor of desert basins far beyond the base of the bordering mountain range. Often, huge monoliths rafted out beyond the mountains by swift-flowing mudflows are stranded on the desert floor. Heavy rains falling onto loose pyroclastic material on the flanks of volcanoes also produce mudflows.

Lahars are mudflows produced by volcanic eruptions that often cause more devastation than the volcano itself. They are a recurring threat at hundreds of volcanoes worldwide. Lahars derive their name from a Javanese word meaning "mudflow" due to their common occurrence in this region. A tragic example was the 1919 eruption of Kelut Volcano on Java, which blew out the crater lake at its summit and created a large mudflow that killed 5,000 people. Lahars are masses of water-saturated rock debris that descend the steep slopes of volcanoes similar to the flowage of wet concrete. Because of their immense power, lahars can carry boulders the size of trucks and travel faster than ordinary floodwaters.

Lahars can be either hot or cold, depending on whether hot rock debris is present. The debris is commonly derived from loose, unstable rock deposited onto the volcano by explosive eruptions. The water is provided by rain, melting snow, a crater lake, or a reservoir next to the volcano. Lahars can also be initiated by a pyroclastic or lava flow moving across a glacier and rapidly melting it. This happened during the May 18, 1980, eruption of Mount St. Helens, which created many destructive mudflows (Fig. 71).

A lahar's speed depends mostly on its fluidity and the slope of the terrain. Lahars can travel swiftly down valley floors for a distance of up to 50 miles or more at speeds approaching 60 miles per hour. Lahars might even travel greater distances than pyroclastic flows and, under favorable conditions, can run as far as 60 miles or more from their sources. Lava flows extending onto glacial ice or snowfields produce floods as well as lahars. Flood hazard zones extend long distances down some valleys. For volcanoes in the western Cascade Range, flood hazard zones reach as far as the Pacific Ocean. Losses from lahars decrease rapidly with increasing height above the valley floor and decrease gradually with increasing distance from the volcano. The vast carrying power of lahars can easily sweep away people and their homes.

The most tragic example of a mudflow initiated by a volcanic eruption in recent history was the November 13, 1985, eruption of Nevado del Ruiz in Colombia. The eruption melted the volcano's ice cap and sent floods and mudflows cascading down the mountainside into the nearby Lagunilla and Chinchina River valleys. The mudflow had a consistency of mixed concrete and carried off everything in its path. It buried almost all the city of Armero

Figure 71 *A mudflow that destroyed the Highway 504 bridge below the confluence of the North and South Forks Toutle River, from the May 18, 1980, eruption of Mount St. Helens.*

(Photo by J. Cummans, courtesy USGS)

30 miles from the volcano and badly damaged 13 smaller towns, killing a total of 25,000 people and leaving 60,000 homeless.

SUBMARINE SLIDES

Some of the largest and most damaging slides occur on the ocean floor, and 40 giant submarine slides have been located around United States territory. The constant tumbling of seafloor sediments down steep banks churn the ocean bottom into a murky mire. Submarine slides moving down steep continental slopes have buried undersea telephone cables under a thick layer of rubble. Sediments eroding out from beneath the cable leave it dangling between uneroded areas of the seabed, causing the cable to fail. A modern slide that broke submarine cables near Grand Banks, south of Newfoundland, moved downslope at a speed of about 50 miles per hour. During the 1964 Good Friday Alaskan earthquake, submarine slides carried away large sections of the port facilities at Whittier, Valdez, and Seward (Fig. 72).

Figure 72 *Damage to port facilities at Seward, Alaska, from submarine slides generated by the March 27, 1964, Alaskan earthquake.*
(Courtesy USGS)

Submarine flow failures can generate large tsunamis that overrun parts of the coast. For example, in 1929, an earthquake on the coast of Newfoundland set off a large undersea slide that triggered a tsunami, killing 27 people. On July 3, 1992, what appeared to be a large undersea slide sent a 25-mile-long, 18-foot-high wave crashing down on Daytona Beach, Florida, overturning automobiles and injuring 75 people.

On July 17, 1998, a train of three giant waves 50 feet high swept away 2,200 residents of Papua New Guina. The disaster was originally blamed on a nearby undersea earthquake of 7.1 magnitude. However, this was much too small to heave up waves to such heights. Evidence collected during marine surveys of the coast implicated a submarine slide or slump of underwater sediment large enough to spawn the waves. The continental slope bears a thick carpet of sediments, which in places has slid downhill in rapid landslides and slower-moving slumps. The evidence on the ocean floor suggests that large tsunamis can be generated by moderate earthquakes when accompanied by landsliding. This phenomenon makes the hazard much more dangerous than was once thought.

Undersea slides carve out deep submarine canyons in continental slopes. The slides consist of sediment-laden water that is considerably denser than the surrounding seawater, allowing sediments to move swiftly along the ocean floor. These muddy waters, called turbidity currents, can move down the gentlest slopes and transport immensely large blocks. Turbidity currents are also initiated by river discharge, coastal storms, or other currents. They deposit huge amounts of sediment that build up the continental slopes and the smooth ocean bottom below.

The continental slopes incline as much as 60 to 70 degrees and plunge downward for thousands of feet. Sediments that reach the edge of the continental shelf slide off the continental slope by the pull of gravity. Huge masses of sediment cascade down the continental slope by gravity slides that can gouge out steep submarine canyons and deposit great heaps of sediment. They are often as catastrophic as terrestrial slides and can move massive quantities of sediment downslope in a matter of hours.

The submerged deposits near the base of the main island of Hawaii rank among the largest landslides on Earth. On the southeast coast of Hawaii, on Kilauea Volcano's south flank, about 1,200 cubic miles of rock are slumping toward the sea at a geologic breakneck speed of 4 inches or more per year (Fig. 73). It is the biggest thing on Earth that is moving in this fashion. Six miles below the volcano lies a nearly horizontal fault that is slipping at a rate of 10 inches per year, making it the fastest-moving fault in the world. Ultimately, some sort of failure will occur, far more destructive than any of the volcano's eruptions.

Gigantic slices of volcanoes have broken off the Hawaiian Islands and skidded across the ocean bottom, sometimes creating towering tsunamis that

Figure 73 *A stepped topography by subsidence of large landslide blocks on the south flank of Kilauea Volcano, Hawaii.*

(Courtesy USGS/Hawaiian Volcano Observatory)

break on nearby shores. On the island of Kauai, a volcano built up the western portion of the island and then collapsed along its eastern edge in a giant landslide. Later, a new volcano grew in its place, only to collapse as well. Remnants of these massive landslides litter the seafloor around the island.

By far, the largest example of an undersea rockslide from a Hawaiian volcano measured roughly 1,000 cubic miles in size and spread some 125 miles from its point of origin. The collapse of the island of Oahu sent debris 150 miles across the deep-ocean floor, churning the sea into gargantuan waves. When part of Mauna Loa Volcano collapsed and fell into the sea around 100,000 years ago, it created a tsunami 1,200 feet high that was not only catastrophic for Hawaii but might have even caused damage along the coast of California.

At the bottom of the rift valley of the Mid-Atlantic Ridge in the Middle Atlantic lies the remains of a massive undersea avalanche in 10,000 feet of water that surpasses in size any landslide in recorded history. One side of the submarine volcanic mountain range apparently gave way and slid downhill at a tremendous speed, running up and over a smaller ridge farther downslope in a manner of minutes. The slide carried about 4.5 cubic miles of rock debris, which is six times greater than the 1980 Mount St. Helens landslide, the largest in recorded history. The slide probably occurred around half a million years ago and possibly created a tsunami 2,000 feet high that pounced onto Atlantic shores.

SOIL EROSION

Perhaps the greatest limitation to further human population growth is soil erosion (Fig. 74) because it forces the world's farmers to feed an ever-growing population on less topsoil. Soil is one of the most endangered resources. Not only does soil erosion rob fields of precious nutrients, it also increases farming costs and lowers food production. As much as one-third of the global cropland is losing soil at a rate that is undermining any long-term agricultural productivity. In other words, people are "mining" the world's soils faster than nature is putting the earth back.

Prior to the advent of agriculture, natural soil erosion rates were probably no more than 10 billion tons per year, slow enough for new soil to be generated in its place. However, present soil erosion rates are estimated at about 20 billion tons per year, equivalent to the loss of some 15 million acres of arable land. In other words, soil is being lost at twice the rate it is being replenished. Thus, world food production per capita could eventually fall if the loss of topsoil continues and the human population keeps rising at staggering levels.

Figure 74 *Severe gully erosion on a pasture in Shelby County, Tennessee, as a result of cultivating unsuited soil that should have been left fallow.*

(Photo by Tim McCabe, courtesy USDA Soil Conservation Service)

Falling rain erodes surface material by impact and runoff. Impact erosion is most effective in regions with little or no vegetative cover and subjected to sudden downpours such as desert areas. The impact of raindrops striking the ground with a high velocity loosens material and splashes it up into the air. On hillsides, some of this material falls back lower down the slope. About 90 percent of the energy dissipates by the impact. Most of the impact splashes rise to about a foot, with the lateral splash movement about four times the height.

Splash erosion accounts for the puzzling removal of soil from hilltops where little runoff occurs. It also ruins soil by splashing up the light clay particles, which are carried away by runoff, leaving infertile sand and silt behind. Rainwater not infiltrating into the ground runs down the hillside and erodes the soil, cutting deep gullies into the terrain. The degree of erosion depends on the steepness of the slope and the type and amount of vegetative cover.

Soil erosion rates vary depending on the amount of precipitation, the topography of the land, the type of rock and soil (Table 4), and the amount of vegetative cover. Efforts to increase worldwide crop production through deforestation, wetland drainage, irrigation, use of artificial fertilizers, genetic engineering, and other scientific methods could ultimately fail if, in the long run, the topsoil disappears. Today's high population growth requires a nearly 2 percent annual increase in food production to meet the global demand. Most of this increase must be made by new technology, especially since a significant portion of the world's cropland is already lost.

TABLE 4 SUMMARY OF SOIL TYPES

Climate	**Temperate (> 160 in. rainfall)**	**Temperate (< 160 in. rainfall)**	**Tropical (heavy rainfall)**	**Arctic or Desert**
Vegetation	Forest	Grass and brush	Grass and trees	Almost none, no humus development
Typical area	Eastern U.S.	Western U.S.		
Soil type	Pedalfer	Pedocal	Laterite	
Topsoil	Sandy, light-colored; acid	Enriched in calcite; white color	Enriched in iron and aluminum, brick red color	No real soil forms because no organic material; chemical weathering very low
Subsoil	Enriched in aluminum, iron, and clay; brown color	Enriched in calcite; white color	All other elements removed by leaching	
Remarks	Extreme development in conifer forests, abundant humus makes groundwater acid; soil light gray due to lack of iron	Caliche—name applied to accumulation of calcite	Apparently bacteria destroy humus, no acid available to remove iron	

Many rivers, particularly those in Africa, which has the worst erosion problem in the world, are becoming heavily sedimented due to topsoil erosion. The Mississippi River draining America's heartland dumps millions of tons of sediment eroded off Midwest farms into the Gulf of Mexico each year. Eroding cropland is costing the United States nearly a billion dollars annually because of polluted and sedimented rivers and lakes. The sediments also severely limit the life expectancy of dams built for water projects such as irrigation. Therefore, the best way to control sediment buildup in reservoirs is by adopting effective soil conservation measures in the watershed so that less topsoil is lost to erosion.

The soil profile (Fig. 75) begins with the A zone, which contains most of the soil nutrients. It is a thin bed from a few inches to a few feet thick, with an average thickness of 7 inches worldwide. Below lies the B zone, which is coarser and of poorer soil quality. As the A zone thins out and erosion brings the B zone to the surface, the potential for runoff and erosion is significantly increased because the B zone is generally unfavorable for sustaining vegetation, whose roots are needed to hold the soil in place.

To keep up with an ever-growing human population, which adds another 100 million mouths to feed each year, farmers have abandoned sound soil conservation practices in favor of more intensified farming methods. This includes less rotation of crops, greater reliance on row crops, more plantings between fallow periods, and extensive use of pesticides and chemical fertilizers rather than natural organic fertilizers that help bind the soil. Three times as much food must be grown on the existing land to meet the global demand of a doubling of the human population expected for the middle of the century.

Over the last 150 years, intensive agriculture has reduced the average soil depth in the United States by about half. The nation's cropland has shrunk by 7 percent since 1980. Presently, upward of 5 billion tons of topsoil are lost annually. Because of rapid urbanization, 1 percent or more of the most productive farmland is lost annually. Expected rises in temperatures, increased evaporation rates, and changes in rainfall patterns brought on by global climate change could further weaken the nation's ability to grow enough food for its own consumption. The production of excess food for export might therefore be severely restricted, leading to mass starvation in countries that have already ruined their land and cannot adequately feed themselves without outside aid.

Throughout the world, most of the arable land is already under cultivation. Efforts to cultivate substandard soils are leading to poor productivity and ultimately abandonment, which in turn leads to severe soil erosion. Marginal lands, which are often hilly, dry, or contain only thin, fragile topsoils and therefore erode easily, are also forced into production. As world populations continue to grow geometrically on a planet whose resources are dwindling rapidly, the vast majority of people could face a cultivation catastrophe.

After discussing landslides, rockslides, soil slides, submarine slides, and related phenomena, including mudflows and erosion, the next chapter deals with other forms of ground movement caused by geologic collapse—ground failures, subsidence, and volcanic calderas.

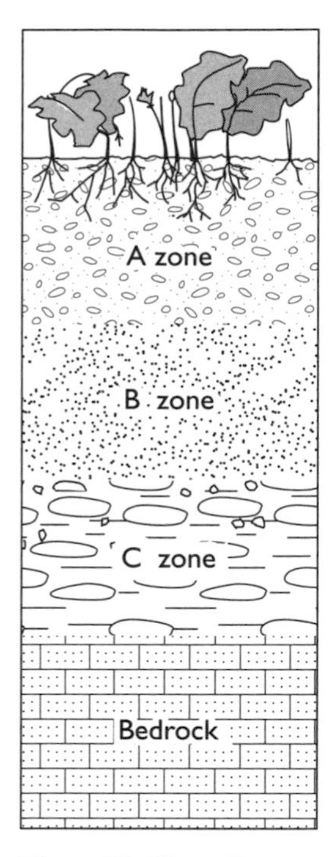

Figure 75 *The soil profile. A zone—organic rich, B zone—organic poor, C zone—parent rock material.*

5

CATASTROPHIC COLLAPSE

THE SINKING OF THE LAND

This chapter examines the geologic processes that cause the earth to collapse. The surface of the Earth is pockmarked by various geologic structures resulting from catastrophic ground failures. They occur when subterranean sediments liquefy during earthquakes or violent volcanic eruptions, causing considerable damage to buildings and other structures. The weakening of sediment layers due to earthquakes can also cause massive subsidence. Large earthquakes whose faults cut the surface slice up the ground, producing large breaks in the crust called fissures. In addition, earthquakes produce tall scarps, where one block sinks below another, which radically transform the landscape. Catastrophic collapse is perhaps best demonstrated at volcanic calderas. These form when the roof of a magma chamber collapses or when a volcano blows off its peak, leaving a broad depression or caldera.

Human activity has produced extensive and alarming geologic effects. Among the most far-reaching is the sinking of the land due to subsidence. Such ground failures result from the dissolution of soluble materials underground or the withdrawal of fluids from subsurface sediments, leading to subsidence, or horizontal depression of the surface. This growing problem is worsening as people draw ever more deeply on stocks of water and petroleum.

THE SINKING EARTH

The 1811 and 1812 New Madrid, Missouri, earthquakes, three of the greatest temblors in American history, caused major changes in ground levels over large areas. The town itself was totally demolished and subsided more than 12 feet. New lakes littered with drowned cypress trees filled the basins of down-dropped crust. Collapsing riverbanks changed the course of the Mississippi River toward the west. Whole islands disappeared, while new ones emerged elsewhere.

The saturated bottomland soil spurted huge geysers of sand and black water 100 or more feet into the air, forming craters up to 30 feet wide. Subsurface sand and water were ejected to the surface, leaving underground voids that caused compaction of subterranean materials and ground settling. The fountaining of subsurface sediment was so extensive that sprayed sand is still visible against the dark soil in Arkansas and Missouri.

During the April 18, 1906, San Francisco, California, earthquake, the ground subsided several inches under buildings, causing them to collapse (Fig. 76). The damage was most severe in the low-lying business district, where the earthquake destroyed or structurally weakened almost all buildings in the downtown area. The subsidence ruptured water mains, and firefighters stood by helplessly as most of the city burned to the ground. Buildings that managed to survive the earthquake were utterly destroyed by fire. Therefore, the ground failure of sediments underneath San Francisco was largely responsible for the city's destruction.

The March 27, 1964, Good Friday Alaskan earthquake was the largest recorded on the North American continent. The estimated area of destruction was 50,000 square miles. Massive landslides and subsidence devastated large sections of Anchorage (Fig. 77). Thirty blocks were destroyed when the city's slippery clay substratum slid toward the sea. Landslides caused significant damage, and houses were destroyed when 200 acres were carried seaward. The ground beneath the port cities of Valdez, Seward, and Whittier gave way, causing large sections to slide toward the sea.

Niigata, Japan, in the northwestern main island of Honshu, overlies a large deposit of natural gas dissolved in saltwater. Huge quantities of water had to be pumped out of the ground to extract the gas, producing an alarming amount of subsidence. Parts of the city sank below sea level, requiring the construction of dikes to keep the sea out. On June 16, 1964, a major earthquake struck the area. It breached the dikes, and the city subsided a foot or more, resulting in serious flooding in the area of subsidence.

Figure 76 *The April 18, 1906, San Francisco, California, earthquake caused these houses to shift to the left, while the tall house dropped from its south foundation wall and leaned against its neighbor.*

(Photo by G. K. Gilbert, courtesy USGS)

In Mexico City, overpumping of groundwater caused the Western Hemisphere's largest metropolis to subside more than 20 feet since 1940. The discharge of water wells far exceeds the natural recharge of the aquifer beneath the city, reducing the fluid pressure as the water table lowers. Some areas subsided more than a foot a year, often resulting in numerous earth tremors. The constant shaking of the ground might explain why residents ignored the foreshocks that preceded the destructive September 19, 1985, earthquake that destroyed large portions of the city and killed as many as 10,000 people.

Collapsing volcanoes throughout the world have brought massive death and destruction. Indeed, the catastrophic collapse of an unstable cone is considered a normal event in the life cycle of a volcano. When the Thera Volcano erupted in the early 15th century B.C., the final colossal explosion caused the cone to collapse into the emptied magma chamber beneath the island, forming a deep, water-filled crater. The collapse of Thera created an immense sea wave, or tsunami, hundreds of feet high that battered the shores and harbors of Crete and other areas in the eastern Mediterranean.

The collapse of Mount Tambora on April 11, 1815, formed a gaping caldera and initiated a massive landslide. Hot pyroclastic flows cascaded down the mountain, wiped out the town of Tambora, and created powerful air currents that uprooted trees miles away. More than 80,000 people perished in the

Figure 77 *The Turnagain Heights landslide from the March 27, 1964, Alaskan earthquake.*
(Courtesy USGS)

disaster. Similarly, on August 27, 1883, practically the entire island of Krakatoa disappeared with a mighty convulsion. After the explosion, the crust of the earth caved into the emptied chamber, creating a large caldera 1,000 feet deep with only its jagged edges protruding above the surface. Most of the damage and the loss of 36,000 lives resulted from huge tsunamis, some more than 100 feet high.

GROUND FAILURES

Ground failures in water-saturated subterranean sediments during violent earthquakes and volcanic eruptions result from a process known as liquefaction. It generally occurs during earthquakes of magnitude 6 or greater and where the ground is unstable and tends to flow when stressed. The potential for disaster is enormous because many of the world's major cities are partly built on young sediments that are subject to liquefaction during earthquakes.

Certain geologic and hydrologic environments are prone to liquefaction. This occurs particularly in areas where sediments were deposited since the last Ice Age, or during the last 10,000 years, and where groundwater lies near the surface, usually within 30 feet. Generally, the younger and less consolidated the sediments and the shallower the water table, the more susceptible the soil is to liquefaction. Clay-free soils, primarily sands and silts, temporarily lose strength and behave as viscous fluids rather than as solid materials.

When seismic waves from an earthquake pass through a loose, saturated, granular soil layer, they distort the structure and cause void spaces to collapse in loosely packed sediments. Each collapse places stress onto the pore water surrounding the grains, which disrupts the soil and increases pressure of the pore water, causing it to drain. If drainage is restricted, the pore water pressure builds until it equals the pressure exerted by the weight of the overlying ground. Grain contact stress is temporarily lost, and the granular soil layer flows like a fluid.

Sand boils often develop during the liquefaction process. They are fountains of water and sediment that spout upward of 100 feet or more from the pressurized liquefied zone. Earthquakes can turn a solid, water-saturated bed of sand underlying less permeable surface layers into a pool of pressurized liquid that seeks a channel to the surface (Fig. 78). Water laden with sediment vents to the surface by artesianlike water pressures developed during liquefaction. Sand boils can also cause local flooding and the accumulation of large deposits of sediment (Fig. 79). The expulsion of sediment-laden fluids from below ground also forms a large subsurface cavity that causes the overlying layers to subside.

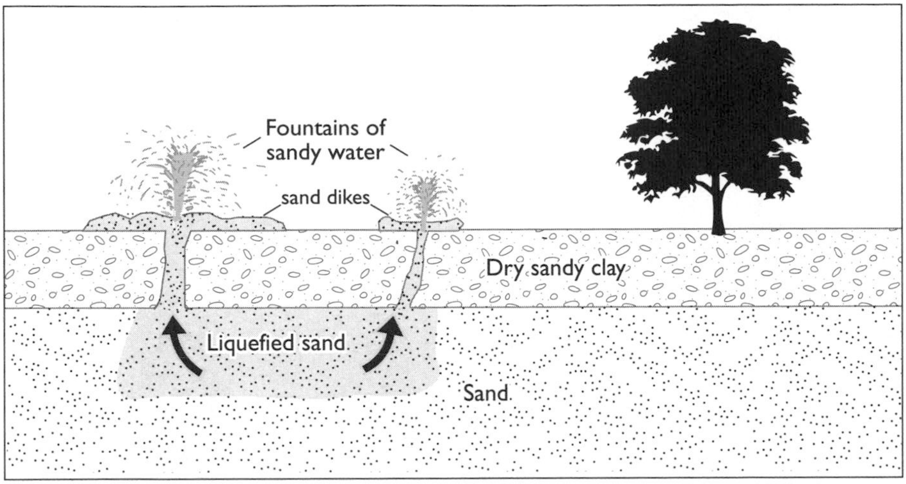

Fountains of sandy water
sand dikes
Dry sandy clay
Liquefied sand
Sand

Figure 78 *Sand boils are fountains of water and sediment that spout from a pressurized, liquefied zone during earthquakes.*

Ground failures associated with liquefaction include lateral spreads, flow failures, and loss of bearing strength. Lateral spreads (Fig. 80) are lateral movements of large blocks of soil in a subsurface layer during earthquakes. They usually break up internally, forming fissures and scarps, and generally develop on gentle slopes of less than 6 percent grade. Horizontal movements on lateral spreads are as great as 10 to 15 feet. However, where slopes are particu-

Figure 79 *Sand boils in an irrigated field near Hollister from the October 17, 1989, Loma Prieta, California, earthquake.*

(Photo by G. Plafker, courtesy USGS)

Figure 80 *Lateral spread on Heber Road, showing associated displacements, fissures, and sand boils, from the October 15, 1979, Imperial Valley, California, earthquake.*

(Photo by C. E. Johnson, courtesy USGS)

larly favorable and the duration of the earthquake is long, lateral movement can extend up to 100 feet or more.

During Alaska's 1964 Good Friday earthquake, lateral spreading of floodplain deposits near river channels damaged or destroyed more than 200 bridges. The lateral spreads compressed bridges over the channels, buckled decks, thrust sedimentary beds over abutments, and shifted and tilted abutments and piers. Lateral spreads are also destructive to underground structures such as pipelines. During the 1906 San Francisco earthquake, several major water main breaks hampered fire-fighting efforts. The inconspicuous ground failure displacements of up to 7 feet were largely responsible for the destruction of San Francisco (Fig. 81). To prevent this occurrence in the future, the city was rebuilt with duplicate water mains so that if one line ruptured during an earthquake, water could be shut off and rerouted to another.

Flow failures are the most catastrophic type of ground failure associated with liquefaction. They consist of soil or blocks of intact material riding on a layer of liquefied sediments, usually over a distance of several tens of feet. However, under certain geographic conditions, they can travel at great speeds over distances of up to several miles. Flow failures usually develop in loose, saturated sands and silts on slopes with greater than 6 percent grade and originate on land and on the seafloor near coastal areas.

The 1920 Kansu, China, earthquake induced several massive flow failures that killed as many as 180,000 people. The largest and most damaging flow failures have taken place undersea in coastal areas. The 1964 Good Friday earthquake in Alaska produced submarine flow failures that destroyed seaport facilities. The flow failures also generated large tsunamis that overran coastal areas and caused additional damage and casualties.

When the ground supporting buildings and other structures liquefies and loses bearing strength, large deformations occur within the soil, causing settlement or collapse. Soils that liquefy under buildings induce bearing failures that cause the structures to subside or tip over. Deformation usually occurs whenever a layer of saturated, cohesionless sand or silt extends from near the surface to a depth of about the width of the building. The most spectacular example of this type of ground failure occurred during the June 16, 1964, Niigata, Japan, earthquake when several multistory apartment buildings tilted as much as 60 degrees (Fig. 82). Most of the buildings were jacked back into an upright position and underpinned with piles to prevent recurring ground failures.

Earthquakes can also cause certain clays, called quick clays, to lose strength and fail. Quick clay is composed of flakes of clay minerals arranged in very fine layers and has a water content of 50 percent or more. Under ordinary conditions, quick clay is a solid that can support a weight of more than a ton per square foot of surface area. However, the slightest jarring from an earthquake can immediately turn it into a liquid. The large landslides in Anchorage, Alaska, during the 1964 Good Friday earthquake resulted from the failure of layers of quick clay along with other beds of saturated sand and silt.

Figure 81 *San Francisco in flames after the April 18, 1906, earthquake due largely to the unavailability of water for fire-fighting efforts caused by broken water mains.*

(Photo by T. L. Youd, courtesy USGS)

Figure 82 *Apartment buildings in Niigata, Japan, that tipped because the loss of bearing strength caused by liquefaction in the underlying sediments during the 1964 Niigata earthquake.*

(Courtesy USGS)

The severity of the earthquake was responsible for the loss of strength in the clay layers and liquefaction in the sand and silt layers, which were the major causes of ground failures that destroyed much of the city (Fig. 83).

Severe subsidence also occurs in permafrost regions of the higher latitudes. Solifluction is the slow downslope movement of waterlogged sediments that causes ground failures in colder climates. When frozen ground melts from the top down during spring in the temperate regions or during summer in permafrost regions, it causes the soil to glide downslope over a frozen base. Solifluction can create many construction problems, especially in areas of permafrost. Foundations must extend down to the permanently frozen layers, or entire buildings might be damaged by the loss of support or by lateral movement downslope.

Another type of movement of soil material is called frost heaving. It is associated with cycles of freezing and thawing mainly in the temperate climates. Frost heaving thrusts boulders upward through the soil by a pull from above and by a push from below. If the top of the rock freezes first, it is pulled upward by the expanding frozen soil. When the soil thaws, sediment gathers below the rock, which settles at a slightly higher level. The expanding frozen soil lying below also heaves the rock upward. After several frost-thaw cycles, the boulder finally comes to rest on the surface, a major annoyance to northern farmers who, every spring, find a new crop of rocks in their fields. Rocks have also been known to push through highway pavement, and fence posts have been shoved completely out of the ground by frost heaving.

Another type of frost action can produce mechanical weathering by exerting pressures against the sides of cracks and crevices in rocks when water freezes inside them, resulting in frost wedging. This widens the cracks, while surface weathering rounds off the edges and corners. This provides a landscape resembling numerous miniature canyons up to several feet wide carved into solid bedrock.

SUBSIDENCE

Subsidence is the downward settling or collapse of the land surface either locally or over broad regional areas without appreciable horizontal movement. It is a common problem throughout the world and is mostly caused by the withdrawal of fluids or by shock waves from earthquakes. Coastal subsidence brought on by earthquakes causes vegetated lowlands that are adequately ele-

Figure 83 *The east part of the Turnagain slide in Anchorage from the March 27, 1964, Alaskan earthquake.* (Courtesy USGS)

Figure 84 *Reelfoot Lake, Tennessee, created by flooding down-dropped crust from the 1811–1812 New Madrid, Missouri, earthquake.*

(Photo by M. L. Fuller, courtesy USGS)

vated to avoid being inundated by the sea to sink far enough to be submerged regularly and become barren tidal mudflats. Between great earthquakes, sediments fill the tidal flats and raise them to the level where vegetation can again grow. Therefore, repeated earthquakes produce alternating layers of lowland soil and tidal flat mud.

Earthquake-induced subsidence in the United States has occurred mainly in California, Alaska, and Hawaii. The subsidence results from vertical displacements along faults that can affect broad areas. The town of New Madrid, Missouri, was totally demolished when the ground beneath it collapsed from a height of 25 feet to 12 feet above the level of the Mississippi River during massive earthquakes in the winter of 1811–1812. Lakes formed in the basins of the down-dropped crust, the largest of which is the 50-foot-deep Reelfoot Lake (Fig. 84). The expulsion of sediment-laden fluids from below ground forms a large subsurface cavity that causes the overlying layers to subside. The New Madrid earthquakes caused subsurface water and sand to spout to the surface, leaving void spaces in the ground that caused compaction of subterranean materials and subsidence.

During the 1906 San Francisco earthquake, the ground subsided several inches, causing settlement and lateral movement in subsurface materials. Roadbeds sank and spread sideways, causing the pavement to split. Roadways buckled and heaved, leaving streetcar tracks bent and suspended in the air (Fig. 85). During the 1964 Good Friday Alaskan earthquake, subsidence destroyed large sections of Anchorage when the clay substratum slid out toward the sea. The earthquake also forced more than 70,000 square miles of land to tilt downward 3 feet or more, causing extensive flooding in coastal areas of southern Alaska.

Many parts of the world have been steadily sinking due to the withdrawal of large quantities of groundwater or petroleum. Generally, the amount of subsidence is on the order of one foot for every 20 to 30 feet of lowered water table. Underground fluids fill intergranular spaces and support sediment grains. The removal of large volumes of fluid, such as water or petroleum, results in a loss of grain support, a reduction of intergranular void spaces, and

Figure 85 *Buckled streetcar railway tracks due to soil settlement and movement during the 1906 San Francisco earthquake.*

(Photo by G. K. Gilbert, courtesy USGS)

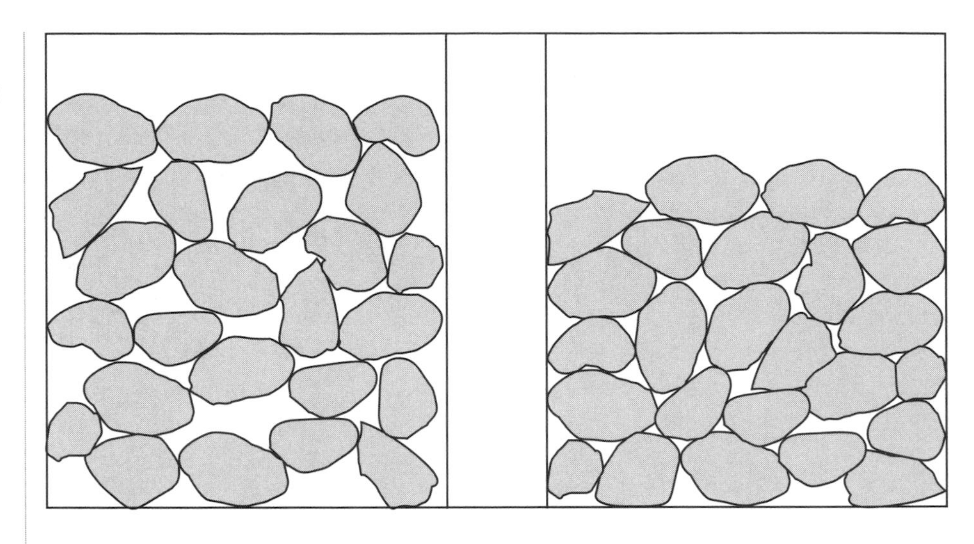

Figure 86 The subsidence of sediments (right) *by the withdrawal of fluids, which causes compaction.*

the compaction of clays. This action causes the land surface to subside wherever widespread subsurface compaction occurs (Fig. 86).

The extraction of natural gas or oil from the ground can also cause earthquakes. Low-magnitude, shallow earthquakes have occurred near gas and oil fields in the United States since the 1920s. The tremors rarely exceed magnitude 4, but they can buckle pipelines and shear off wellheads, resulting in oil spills. Pumping oil out of the ground contracts the rock in the reservoir and sets up large pressure changes over short distances. Vertical contraction pulls down on the overlying rock and causes the ground above the reservoir to sink. Horizontal stresses pull the surrounding rock inward, causing the reservoir to contract like a drying sponge. If the pull is strong enough to shear the rock, it can trigger a mild earthquake. Furthermore, pumping petroleum from the ground can collapse the oil-filled fractures, which seals the remaining reservoir and causes oil wells to go dry.

The opposite condition occurs by pumping waste fluids into wells, which triggers earth tremors. On January 31, 1986, a moderate earthquake struck 25 miles northeast of Cleveland, Ohio, the largest ever reported in the area. Since 1974, some 250 million gallons of hazardous liquids were pumped into two 5,900-foot-deep wells. The resulting pressures enlarged fractures in the rock that might have activated a nearby fault. In the 1960s, a series of earthquakes hit Denver, Colorado, due to wastewater injection at the Rocky Mountain Arsenal. The liquids tend to unlock faults, causing rocks with pent-up pressures to slip. Oil forced out of reservoirs using water injection wells can also cause minor earthquakes. Some 300,000 such wells exist in the United States with the potential of producing mild earth tremors.

The most dramatic examples of subsidence in the United States have occurred along the Gulf Coast of Texas, in Arizona, and in California. The Houston–Galveston, Texas, area has experienced local subsidence of as much as 7.5 feet and widespread subsidence of a foot or more over an area of 2,500 square miles, due mostly to the withdrawal of large amounts of groundwater. In Galveston Bay, the ground subsided 3 feet or more over an area of several square miles due to the rapid pumping of oil from the underlying strata. Subsidence in some coastal towns has increased their susceptibility to flooding during severe coastal storms by dropping them closer to sea level.

At Long Beach, California, the ground subsided, forming a huge bowl up to 26 feet deep over an area of 22 square miles due to the withdrawal of large quantities of oil during the 1940s and 1950s. In some parts of the oil field, the affected land subsided at a rate of 2 feet per year. In the downtown area, the subsidence amounted to upward of 6 feet, causing considerable damage to the city's infrastructure. Most of the subsidence was halted by injecting seawater under high pressure into the underground reservoir, which fortuitously increased the production of the oil field by forcing the petroleum toward the surface.

Large areas of California's San Joaquin Valley have subsided because of intense pumping of groundwater for agricultural purposes. The arid region is so dependent on groundwater it accounts for about one-fifth of all well water pumped in the United States. The ground has been sinking at rates of up to a foot a year. In some areas, the land has fallen more than 20 feet below former levels. Other parts of the San Joaquin Valley have subsided due to shrinking of the soil as it dries out. In the northern part of the valley, subsidence has dropped the land surface more than 10 feet below sea level, requiring the construction of protective dikes to prevent flooding.

Egypt's Nile Delta (Fig. 87) is heavily irrigated and supports 50 million people in a 7,500-square-mile area, about the size of New Jersey. Port Said on the northeast coast of the delta sits at the northern entrance to the Suez Canal and is a bustling seaport of half a million people. The region overlies a large depression filled with 160 feet of mud, suggesting that part of the delta is slowly dropping into the sea. On either side of the 20-mile-wide depression, the mud layer thins to only 40 feet, looking like a drooping clothesline. This indicates that the ancient shoreline took a downward plunge. Over the last 8,500 years, this portion of the fan-shaped delta has been lowering by less than a quarter inch per year. However, more recently, the yearly combined subsidence and sea level rise have greatly exceeded this amount.

The delta is just 3 feet above sea level. A 2- to 3-foot expected rise in sea level by the end of this century could place major portions of the city underwater. Moreover, as the land subsides, seawater infiltrates into the groundwater system, rendering it useless. Dams and artificial channels have

Figure 87 *Egypt's Nile Delta supports about 50 million people in a 7,500-square-mile area.*
(Courtesy NASA)

nearly completely cut off the region's sediment supply from the river, preventing eroded areas along the sea from building back up. In addition, the long-term deposition of heavy, waterlogged sediments is pushing down the underlying crust, a process that commonly occurs with most deltas. Furthermore, two faults border either side of the sinking region, which could lead to catastrophe if an earthquake causes further subsidence.

Venice, Italy, is drowning because of a combination of rising sea levels and subsidence. The city is most unusual because it is built right at the water's edge, where buildings rest half on sea and half on land. Indeed, the entire building site is virtually devoid of dry land. Venice has been battling rising waters since its birth in A.D. 421, with the pace of change accelerating markedly over the last 100 years. The city is built on soft, compactible sediments and is slowly sinking under its own weight. Venice has sunk more than 6 feet since its founding, forcing residents to fill in the lagoon with sand to stay above water.

The people of Venice not only have to fight against the sinking of their city but also against the effects of rising sea levels. The incidence of high tides has increased in magnitude and frequency since 1916. Over the last half of the 20th century, the cumulative subsidence of Venice has been about 5 inches. Meanwhile, the Adriatic Sea has risen about 3.5 inches over the last century, resulting in a change of more than 8 inches between Venice and the sea. The city regularly floods during high tides, heavy spring runoffs, and storm surges.

The more recent subsidence results from the overuse of groundwater, which causes compaction of the aquifer beneath the city. To stop the flooding, locks are needed to keep the sea from spilling into the lagoon upon which the city was built. Unfortunately, such a move could silt up the lagoon, causing Venice's famous canals to go dry. The alternative is to build up the ground level and raise the city as people have done in the past. Either way, Venice would no longer sit at the edge of the sea as it has for centuries, spoiling its major tourist attraction.

In the northeastern section of Tokyo, Japan, the land has sunk around building foundations due to overdrawing of groundwater. The subsidence progressed at a rate of half a foot a year over an area of about 40 square miles, 15 square miles of which sank below sea level. This prompted the construction of dikes to keep out the sea from certain sections of the city. A threat of catastrophe hangs over Tokyo because it could well be inundated by floodwaters during a typhoon or an earthquake. Had the January 17, 1995, Kobe earthquake of 7.2 magnitude struck Tokyo instead, more than half the city would have sunk beneath the waves.

Subsidence from the withdrawal of groundwater can produce fissures, resulting in the formation of open cracks in the ground. Subsidence can also cause the renewal of surface movement in areas cut by faults. Surface fissuring and faulting resulting from the withdrawal of groundwater is a potential problem in the vicinity of Las Vegas, Nevada, as well as in the arid regions of California, Arizona, New Mexico, and Texas. The withdrawal of large volumes of water and oil can cause the ground to subside to considerable depths, often with catastrophic consequences.

Sediments also subside significantly when water is added to them. This condition is especially prevalent in the heavily irrigated dry western states. The land surface has lowered 3 to 6 feet on average and as much as 15 feet in the most extreme cases. The settling occurs when dry surface or subsurface deposits are extensively wetted for the first time since their deposition following the last Ice Age. The wetting causes a reduction in the cohesion between sediment grains, allowing them to move and fill intergranular openings. The compaction produces an uneven land surface, resulting in depressions, cracks, and wavy surfaces.

RESURGENT CALDERAS

If a volcano decapitates itself by blowing off its upper peak (Fig. 88) or collapses into a partially emptied magma chamber, it forms a caldera, a Spanish word meaning "cauldron." Indeed, such a cataclysmic collapse is what keeps volcanoes from becoming any taller. Perhaps the largest caldera in North America is the 45-mile-wide La Garita in southwestern Colorado, which erupted about 28 million years ago. The eruption dwarfs any recent geologic event in both its violence and sheer volume of rock, ash, and lava it spewed across the ancient American West. It lasted several weeks and was thousands of

Figure 88 *Mount St. Helens following the May 18, 1980, eruption, which blew off the upper peak.* (Photo by MSH-Brugman, courtesy USGS)

times larger than the 1980 eruption of Mount St. Helens, devastating an area of 10,000 square miles. It also created Colorado's picturesque San Juan Mountains. Today, the volcano is a crumpled, eroded caldera that straddles a 1,000-square-mile area of the San Juans. Interestingly, the eruption created about a billion dollars worth of gold and silver ores.

A resurgent caldera forms with the sudden ejection of large volumes of molten rock from a magma chamber lying just a few miles beneath the surface. This action abruptly removes the underpinning of the chamber's roof and causes it to collapse, leaving a deep, broad depression on the surface. The infusion of new molten rock into the magma chamber slowly heaves the caldera floor upward, producing a vertical uplift of several hundred feet. Generally, if a large part of the caldera floor bulges rapidly at a rate of several feet a day, a major eruption follows within a few days.

Most younger calderas are hundreds of miles from any subduction zone, where the majority of volcanoes get their supply of new magma. Resurgent calderas usually develop over a mantle plume, or hot spot, that melts near-surface rocks. The calderas generally exist where the crust is thinning and the mantle rises near the surface. Such geologically young calderas are inherently unstable, spurring episodes of unrest not associated with eruption. Generally, caldera eruptions are more the exception than the rule. Nevertheless, calderas are dynamic and in a delicate equilibrium so that even small disturbances can lead to unrest.

The calderas are recognized by widespread secondary volcanic activity, such as hot springs, fumaroles, and geysers. Yellowstone with its world-famous Old Faithful geyser is such a caldera. The region is frequently shaken by earthquakes, the largest of which occurred in 1959 and threw off Old Faithful's dependable timing. In addition, a variety of boiling mud pits and hot-water streams are produced when rainwater seeps into the ground, acquires heat from a magma chamber, and rises through fissures in the torn crust.

The primary requirement for the production of fumaroles and geysers is for a large, slowly cooling magma body to lie near the surface where it provides a continuous supply of heat. The hot water and steam are derived either from juvenile water released directly from magma along with other volatiles or from groundwater that percolates downward near a magma body, where it is heated by convection currents. Volatiles released from the magma body can also heat the groundwater from below.

As with all resurgent calderas, the one at Yellowstone formed above a mantle plume that was large and long lasting enough to melt huge volumes of rock. About 600,000 years ago, a massive volcanic eruption ejected some 250 cubic miles of ash and pumice, creating the Yellowstone Caldera. The caldera's floor has slowly domed upward on average about three-quarters of an inch per year since 1923.

Many other calderas, no more than a few tens of millions of years old, lie in a broad belt covering Nevada, Arizona, Utah, and New Mexico. A million years ago, a massive eruption created the Valles Caldera in northern New Mexico, which has been explored for its geothermal energy potential. Workers from the Los Alamos Scientific Laboratory drilled a well on the flanks of the caldera to a depth of 1.8 miles and encountered temperatures of 200 degrees Celsius. Cold water was pumped down the borehole, and a second recovery well brought the superheated water back to the surface.

The Long Valley Caldera east of Yosemite National Park, California, formed by a cataclysmic eruption 700,000 years ago that resulted in a 20-mile-long, 10-mile-wide, and 2-mile-deep depression (Fig. 89a, b). The eruption fragmented the nearby mountains into rocky debris. Some 140 cubic miles of material were strewn over a wide area as far as the East Coast. The Long Valley Caldera encompasses such features as Devils Postpile National Monument and Mammoth Lakes and is growing about an inch per year.

Magma appears to be moving again into the resurgent Long Valley Caldera from a depth of several miles beneath the surface. The increase in volcanic and seismic activity is indicated by a rise in the center of the caldera's floor of a foot or more since 1980. Several medium-sized earthquakes of magnitude 6 or less have struck the region over the same period. These quakes might indicate that magma is pressing toward the surface and that the caldera is poised for its first eruption in 40,000 years. Mammoth Mountain is a young volcano within the caldera that has experienced an extended period of activity and appears to be ready for eruption, which could flood large portions of neighboring Nevada with thick basalt flows.

Similar eruptions have taken place in other parts of the world within the last million years. In northern Sumatra, the giant Toba Caldera, whose maximum dimension is nearly 60 miles, is the largest resurgent caldera on Earth. It formed 75,000 years ago when a massive eruption, possibly the largest in the last million years, caused the crust to collapse as much as a mile or more.

Other types of calderas form in areas where the crust has been fractured, allowing magma to move toward the surface. The magma intrusion domes the overlying crust upward, creating a shallow magma chamber that contains a large volume of molten rock. The doming process produces stresses in the surface rock that forms the roof of the magma chamber, causing it to weaken and collapse along a ring fracture zone. This becomes the outer wall of the caldera after the eruption.

A smaller caldera results when a powerful volcano erupts and decapitates itself by blowing off its upper peak, leaving a broad crater generally more than a mile across. Often, the lava blows outward, greatly enlarging the crater. The highly viscous lava might also form a plug in the crater, which can slowly rise to form a huge spire or dome.

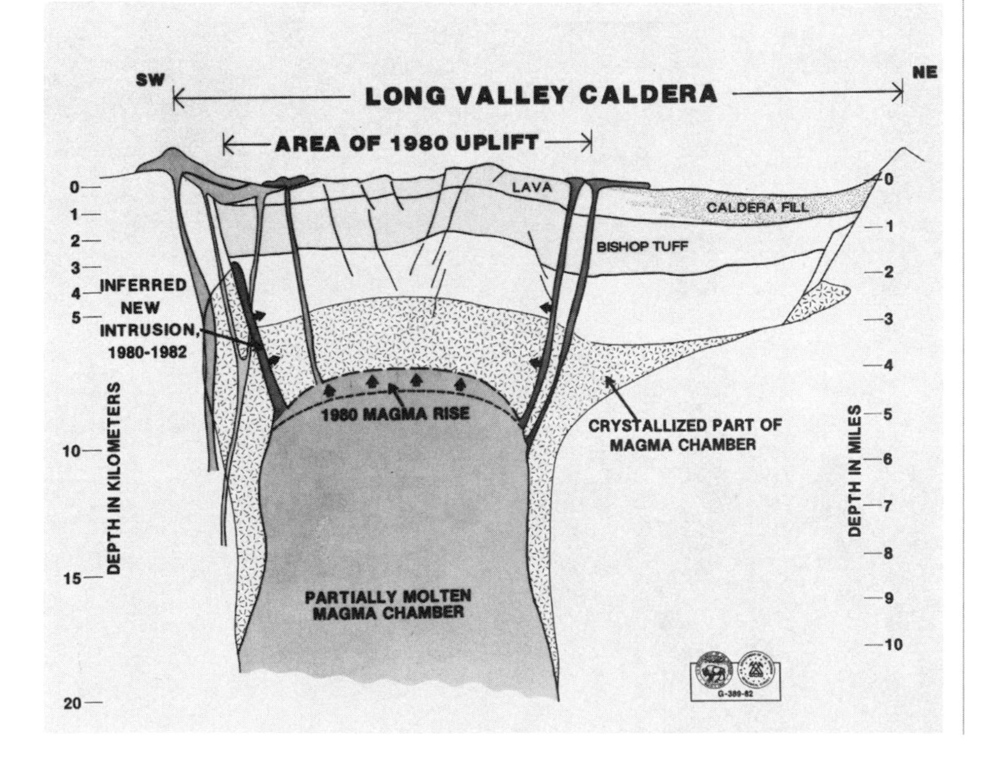

Figure 89a, b *Long Valley Caldera, California.*

(Courtesy USGS)

Dormant calderas, filled with freshwater from melting snow or rain, form crater lakes. They are among the deepest lakes in the world, depending on the depth of the caldera floor and the water level below the crater rim. Erosion widens the caldera, while sediments eroded from the wall accumulate into thick deposits on the lake bottom. Resurgence of the caldera floor sometimes creates an island capped with young lake sediments.

The world's largest crater lake fills the huge Toba Caldera. The caldera formed when the roof over a large magma chamber suddenly collapsed. The floor of the caldera consequently subsided over a mile and filled with water to form a deep lake. Later, the caldera floor heaved upward several hundred feet like a huge piston. This created Samosir Island, which is 25 miles long, 10 miles wide, and possibly still rising.

Oregon's Crater Lake (Fig. 90) originated when the upper 5,000 feet of the 12,000-foot composite cone of Mount Mazama collapsed about 6,000 years ago and filled with rainwater and melted snow. The lake is 6 miles wide and 2,000 feet deep, the sixth deepest in the world. The rim of the caldera rises 500 to 750 feet above the water's surface. At one end of the lake is a small vol-

Figure 90 *Crater Lake, Klammath County, Oregon, created by the collapse of Mount Mazama.*
(Courtesy USGS)

canic peak called Wizard Island, which uplifted during a subsequent period of volcanic activity.

On June 6, 1912, a gigantic explosion of unprecedented violence tore open the bottom of the west slope of Mount Katmai. A mass of 10 cubic miles of pumice, ash, and gas invaded the valley, burning entire forests in its path and filling the valley at some places to a height of 600 feet. For two days, powerful explosions ejected millions of tons of volcanic material from a single fissure 5 miles west of Katmai. The top 1,200 feet of Mount Katmai exploded and collapsed into a giant caldera 1.5 miles wide and 2,000 feet deep. It later filled with water to form a deep crater lake.

COLLAPSE STRUCTURES

Limestone and other soluble rocks underlie large portions of the world. When acidic groundwater percolates through these rocks, it dissolves soluble minerals such as calcite, forming cavities or caverns. Rainwater filtering through overlying sedimentary layers reacts with carbon dioxide to form a weak carbonic acid. This flows downward through cracks in the lower rock layers and dissolves calcite or dolomite. This action enlarges fissures in the rock and eventually creates a path for more acidic water.

Land overlying the caverns can suddenly collapse, forming sinkholes of 100 feet or more in depth and up to several hundred feet across. One of the most dramatic examples of this phenomenon occurred when a sinkhole 350 feet wide and 125 feet deep suddenly opened under Winter Park, Florida, in May 1981, collapsing part of the town. Another impressive example of this phenomenon occurred in Barton, Florida, on May 22, 1967, when a sinkhole 520 feet long and 125 feet wide collapsed under a house (Fig. 91). On December 12, 1995, heavy rainfall and a sewer pipe break in San Francisco, California, created a huge sinkhole as deep as a 10-story building that swallowed a million-dollar house and threatened dozens more.

At other times, the land surface can settle slowly and irregularly. The subsidence can cause extensive damage to buildings and other structures located over the pits formed by dissolving soluble minerals. Although the formation of sinkholes is a natural phenomenon, the process can be accelerated by the withdrawal of groundwater or the disposal of water into the ground.

The landscape formed by numerous sinkholes is called karst terrain. The name is derived from the region of Karst on the coast of Slovenia known for its numerous limestone caves. Karst terrain is generally found in areas with moderate-to-abundant rainfall. Throughout the world, about 15 percent of the land surface rests on such terrain, creating millions of sinkholes.

Figure 91 *A sinkhole 520 feet long, 125 feet wide, and 60 feet deep that collapsed under a house in Barton, Florida.*

(Courtesy USGS)

The major locations of karst terrain in the United States are in the Southeast and Midwest as well as in portions of the Northeast and West. In Alabama, soluble limestone and other sediments cover nearly half the state, where thousands of sinkholes pose serious problems for highways and other construction projects. A third of Florida is underlain by eroded limestone at shallow depths, which is subject to the formation of sinkholes that are often filled with water.

In regions of nearly horizontal limestone strata, flat areas with karst features form karst plains. A blind valley is a river valley in karst terrain that ends abruptly where the stream disappears underground, called a swallow hole. During heavy rains, a blind valley might become a temporary lake. A type of blind valley called a karst valley forms by the coalescence of several sinkholes. Often sinkholes fill with water and become small, permanent lakes.

The jungles of Mexico's Yucatan Peninsula display a bizarre realm of giant undersea caverns and sinkholes of astonishing beauty. The caves are linked by miles of twisting passages 100 feet beneath the ocean. Strange sightless creatures occupying the deepest recesses of the caves are blinded by generations of species living in utter darkness. The sinkholes formed when the upper surface of a limestone formation collapsed, exposing the watery world beneath the jungle floor.

The underground limestone formation is honeycombed with long tunnels, some several miles long, and huge caverns that could easily hold several houses. Like surface caves, the Yucatan caverns contain a profusion of icicle-shaped formations of stalactites hanging from the ceiling and stalagmites clinging to the floor. The formations also include delicate, hollow stalactites called soda straws that took millions of years to create. However, they are often destroyed in mere moments by careless divers exploring the caves.

Blue holes are sinkholes submerged by the sea and appear dark blue because of their great depth. Many blue holes dot the shallow waters surrounding the Bahama Islands southwest of Florida. They formed during the last Ice Age when the ocean dropped several hundred feet, exposing the ocean floor well above sea level. The sea lowered in response to the growing ice sheets that covered the northern regions of the world, locking up huge quantities of water.

During the seabed's exposure as dry land, acid rainwater seeping into the soil dissolved the limestone bedrock, creating vast subterranean caverns. Under the weight of the overlying rocks, the roofs of the caverns collapsed, forming huge gapping pits. At the end of the Ice Age, when the ice sheets melted and the ocean rose to its present level, the area was inundated by the sea and the sinkholes were submerged. Blue holes can be very treacherous places because they often have strong eddy currents or whirlpools that can suck an unwary boat to the bottom during incoming and outgoing tides.

Shallow tunnels in lava flows can collapse, causing the ground to subside. Lava tunnels are long caverns beneath the surface of a lava flow. They are created by the withdrawal of lava as the surface cools and hardens. In exceptional cases, they can extend up to 12 miles inside a lava flow. Often, circular or elliptical depressions exist on the surface of lava flows due to the collapse of lava tunnel roofs (Fig. 92). This phenomenon commonly occurs in the volcanic fields of Alaska, Washington, Oregon, California, and Hawaii. Among the largest is a collapse depression in a lava flow in New Mexico that is nearly a mile long and 300 feet wide.

Sometimes human activity can cause serious surface collapse. The Nevada Test Site, 65 miles northwest of Las Vegas, has taken on the appearance of a moonscape, pockmarked with craters created by underground nuclear tests. The craters formed when subterranean sediments fused into glass by the tremendous heat generated by the explosions. This greatly reduced the volume, causing the overlying sediments to collapse to fill the underground caverns. Sometimes fissures opened on the surface to vent gases escaping from the molten rocks. Many nuclear tests were conducted under the code name Operation Plowshare. Its purpose was to use nuclear detonations for excavating canals, harbors (Fig. 93), mines, and other useful purposes. However, the 1962 ban on atmospheric nuclear detonations prohibited any further testing.

Figure 92 *A collapsed lava tunnel, Craters of the Moon National Monument, Idaho.*

(Photo by H. T. Sterns, courtesy USGS)

Figure 93 *Artist concept of nuclear harbor excavation.*

(Photo courtesy Lawerence Livermore National Laboratory and U.S. Department of Energy)

Figure 94 *Subsidence depressions, pits, and cracks above an abandoned coal mine, Sheridan County, Wyoming.*

(Photo by C. R. Dunrud, courtesy USGS)

Human activity often affects the stability of the surface terrain. By using satellites to detect small movements of the Earth's surface, researchers found a large area of subsidence in Vomero, west of Naples, Italy. The ground subsided as much as 10 inches between 1992 and 1996 during the construction of a new underground railway line. The subsidence gradually slowed after the completion of the Vomero metro tunnel. The satellites have detected subsidence in other locations in the world where underground excavations occur.

The collapse of abandoned underground coal mines, especially those in the eastern United States, might leave the rocks above the mine workings with inadequate support and cause the surface to drop several feet, forming several depressions and pits (Fig. 94). In situ coal gasification and oil shale retorting can also cause the overlying ground to subside. Solution mining uses large volumes of water pumped into the ground to remove soluble minerals, such as salt, gypsum, and potash. It produces huge underground cavities that can collapse and cause surface subsidence. If the mines exist under towns, the overlying buildings might be heavily damaged or destroyed as the underground cavities collapse.

After a discussion of geologic collapse, the next chapter investigates the most common geologic hazard—flooding—including the occurrences of flooding, flood types, flood-prone regions, and flood control.

6

FLOODS

RIVER OVERFLOWS

This chapter examines the effects of flooding on people and their property. Floods are naturally recurring events and important geologic processes that alter the courses of rivers and distribute soils over the land. Heavily sedimented rivers also increase the severity of floods. Whenever a major flood occurs, it often reshapes the landscape through which the river flows. While a flood is in progress, a river might alter its course several times as it rushes to the sea.

Floodplains on either side of a river channel function to carry excess water during floods. They become hazardous only when people occupy flood-prone areas. Failure to recognize this function has led to haphazard development in these areas with a consequent increase in flood dangers (Table 5). Floodplains are a valuable natural resource that must be managed properly to prevent flood damage. Because people insist on building in flood-prone areas without recognizing the purpose of floodplains, floods are becoming increasingly hazardous, killing people and livestock, destroying property, and displacing inhabitants from their homes.

TABLE 5 CHRONOLOGY OF MAJOR U.S. FLOODS

Date	Rivers or Basins	Damage (in $millions)	Death toll
1903	Kansas, Missouri, and Mississippi	40	100
1913	Ohio	150	470
1913	Texas	10	180
1921	Arkansas River	25	120
1921	Texas	20	220
1927	Mississippi River	280	300
1935	Republican & Kansas	20	110
1936	Northeast U.S.	270	110
1937	Ohio & Mississippi	420	140
1938	New England	40	600
1943	Ohio, Mississippi, and Arkansas	170	60
1948	Columbia	100	75
1951	Kansas & Missouri	900	60
1952	Red River	200	10
1955	Northeast U.S.	700	200
1955	Pacific Coast	150	60
1957	Central U.S.	100	20
1964	Pacific Coast	400	40
1965	Mississippi, Missouri, and Red Rivers	180	20
1965	South Platte	400	20
1968	New Jersey	160	—
1969	California	400	20
1969	Midwest	150	—
1969	James	120	150
1971	New Jersey & Pennsylvania	140	—
1972	Black Hills, S. Dakota	160	240
1972	Eastern U.S.	4,000	100
1973	Mississippi	1,150	30
1975	Red River	270	—
1975	New York & Pennsylvania	300	10
1976	Big Thompson Canyon	—	140
1977	Kentucky	400	20

TABLE 5 (CONTINUED)

Date	Rivers or Basins	Damage (in $millions)	Death toll
1977	Johnstown, Pennsylvania	200	75
1978	Los Angeles	100	20
1978	Pearl River	1,000	15
1979	Texas	1,250	—
1980	Arizona & California	500	40
1980	Cowlitz, Washington	2,000	—
1982	Southern California	500	—
1982	Utah	300	—
1983	Southeast U.S.	600	20
1993	Midwest U.S.	12,000	24
1997	Red River, North Dakota	1,000	—
1999	Tar River, North Carolina	6,000	—

HAZARDOUS FLOODS

The worst flood-related disaster in modern history occurred in 1887 when the Yellow River overflowed its levees and flooded much of northern China. The flood drowned 7 million people in what has been labeled as "China's Sorrow." China has the most irrigated land of any country in the world to feed its growing population, which might account for its many killing floods. China's water is provided by some 100,000 dams and reservoirs. They have a total storage capacity of about 100 cubic miles of water. During the 20 years from 1947 to 1967, more than 150,000 people lost their lives to flooding in southern and southeastern Asia. In the same period, some 1,300 lives were lost in the United States during a time of unprecedented floods.

The greatest dam disaster in North America took place on the south fork of the Little Conemaugh River, 13 miles northeast of Johnstown, Pennsylvania. A population of 55,000 people lived within the narrow river valley, with 30,000 living in Johnstown alone. An earthen dam more than 70 feet high and 900 feet across, built in 1852 as part of a canal project, was later abandoned in favor of the railroad. This left the reservoir mostly unused and neglected. Thirty years after the dam was built, a major break was filled in, and a new lake was created to promote sport fishing in the area.

On May 31, 1889, rain began falling in torrents for 36 hours, with a total accumulation of 8 to 10 inches. The heavy spring rains rapidly raised the level of the reservoir, causing the water to flow over the top of the dam and spout from its foundation. The bulging reservoir then pushed aside the weakened dam, and a 40-foot-high wall of water raced down the valley below. One by one, small communities downstream were swept along by the raging waters. Just 15 minutes after the dam break, the floodwaters reached Johnstown.

The flood raced at an incredible speed along the constricted valley as the wall of water made a deep, steady rumble that grew louder and louder as it crashed down on the city. The flood obliterated everything in its path and carried people off to their deaths. Hundreds of houses and other debris piled up in front of a sturdy stone railroad bridge that spanned the river in the center of town. The huge jumble of debris subsequently caught fire. As many as 2,000 people trapped in the wreckage were burned alive. Estimates of the number of people killed within 20 miles downstream of the dam ranged from 7,500 to as many as 15,000. The tragedy was made even more painful because people had been warned of the dam's weakened condition and the disaster could have easily been avoided.

Another major dam break flood occurred at the Teton Dam near Newdale, Idaho, on June 5, 1976. As the Teton Reservoir behind the 130-foot-high earthen dam was being filled, seepage began in the dam wall, which severely eroded the downstream dam embankment. The weakened embankment subsequently fell into the reservoir, which breached the dam as water cascaded into the canyon below (Fig. 95). As the fast-moving floodwaters emerged from the canyon mouth 5 miles downstream, the flood waves spread rapidly over the widening floodplain of the Teton River.

The dam break caused a flood of unprecedented magnitude on the Teton River, lower Henrys Fork, and Snake River. A wall of water up to 16 feet high devastated communities downstream of the dam. The rampaging floodwaters carried off large trees and debris from destroyed buildings and other structures. The water spread over an area of more than 180 square miles, causing damages of about $400 million. Luckily, because of advance flood warnings, only 11 lives were lost.

One of the most disastrous flash floods raged through the Big Thompson River Canyon east of Rocky Mountain National Park in north-central Colorado on July 31, 1976. Thunderstorms in the canyon area dumped some 10 inches of rain in a 90-minute interval. Billions of gallons of water rushed down the steep slopes overlooking the river and poured into the narrow canyon. The river rose with furious speed and set in motion a terrifying flood on the Big Thompson River and its tributaries between Estes Park and Loveland. People vacationing in the canyon fled uphill for their lives as a 20-foot-high wall of water bore down on them.

Figure 95 *The June 5, 1976, Teton Dam break, which caused extensive flooding downstream.* (Courtesy USGS)

For almost the entire 25-mile stretch of the canyon, people scampered for the safety of the high ground ahead of the sudden onslaught of water. The floodwaters carried off buildings, vehicles, and large trees (Fig. 96). The flood destroyed the town of Drake between Estes Park and Loveland. It also washed out almost the entire length of the highway through the canyon, leaving thousands of people stranded. People were evacuated by Army helicopters, while rescue crews using four-wheel-drive vehicles searched the muddy riverbanks for other survivors. In many cases, they found the gored remains of those who did not escape the raging waters. Bodies were disfigured almost beyond recognition from the violent action of the water and from being battered against rocks and other objects. The flood also wiped out several small communities, resulting in more than $35 million in damages. At least 139 lives were lost, and hundreds of people were injured.

Figure 96 *A wrecked house and other debris near Drake, Colorado, from the July 31, 1976, Big Thompson River flood.*

(Courtesy USGS)

The 1980 eruption of Mount St. Helens in southwestern Washington produced major debris flows, mudflows, and severe flooding on the Toutle and lower Cowlitz Rivers. Heavy runoff from melted glaciers and snowfields on the volcano's flanks supplemented by outflows from Spirit Lake below the volcano were the main sources of the floodwaters. Great volumes of sediment and thousands of fallen trees transported during the flood on the Toutle River destroyed most of its bridges (Fig. 97). Much of the sediment carried downstream into the Cowlitz and Columbia Rivers formed a shoal that blocked shipping for several days.

The deadliest natural disaster in American history hit the Galveston, Texas, area on September 8, 1900. The resort town had a population of about 38,000. It was located on the east end of Galveston Island, connected to the mainland by only a single, long bridge. It was a city literally built on sand, with an average elevation of only 6 feet above sea level. When the eye of a hurricane approached the island, it generated wind speeds of more than 110 miles per hour. The seas submerged the bridge, and the people of Galveston lost their only way out.

As a massive storm surge swept through town, it crumpled buildings and sent people into the surging coastal waters where they were taken helplessly out with the tide. When calm finally returned, 10,000 to 12,000 people were found dead. The disaster prompted the construction of a seawall to protect against any future storm. The project paid off because 15 years later, a similar hurricane bore down on Galveston, this time killing fewer than a dozen people.

The crowded Bay of Bengal, Bangladesh, with a population of 100 million crammed into a country about the size of Wisconsin, is frequently pounded by cyclones on the Indian Ocean. The year before Bangladesh achieved its independence from Pakistan, an immense cyclone hit the Bay of Bengal on November 13, 1970, and killed upward of 1 million people, making it one of the world's worst natural disasters in history. On May 24, 1985, Bangladesh suffered its 60th killer cyclone since 1822. The cyclone whipped up winds of more than 100 miles per hour. The winds pushed up a 15- to 50-foot tidal wave that swept over a cluster of islands in the shallow bay. When the storm was done, upward of 100,000 people were dead and 250,000 others were homeless. The storm left 30,000 cattle dead, 3,000 square miles of cropland ravaged, and vital fishing grounds wasted. When rescue workers finally reached the islands, they found whole settlements along with fishing boats swamped or washed into the sea.

During the winter of 1982–1983, powerful storms brought destructive winds, tides, floods, and landslides to the California coast that caused more than $300 million in damages and forced 10,000 people to evacuate their homes. As the storms marched eastward, they drenched the southwestern states and placed many parts of America's southland underwater, requiring the evacuation of tens of thousands of people from their homes. A huge snowpack in the Colorado Rockies brought the Colorado River to flood stage. The

Figure 97 *Destruction of the St. Helens bridge caused by a mudflow from the 1980 Mount St. Helens eruption.*

(Photo by MSH-Schuster, courtesy USGS)

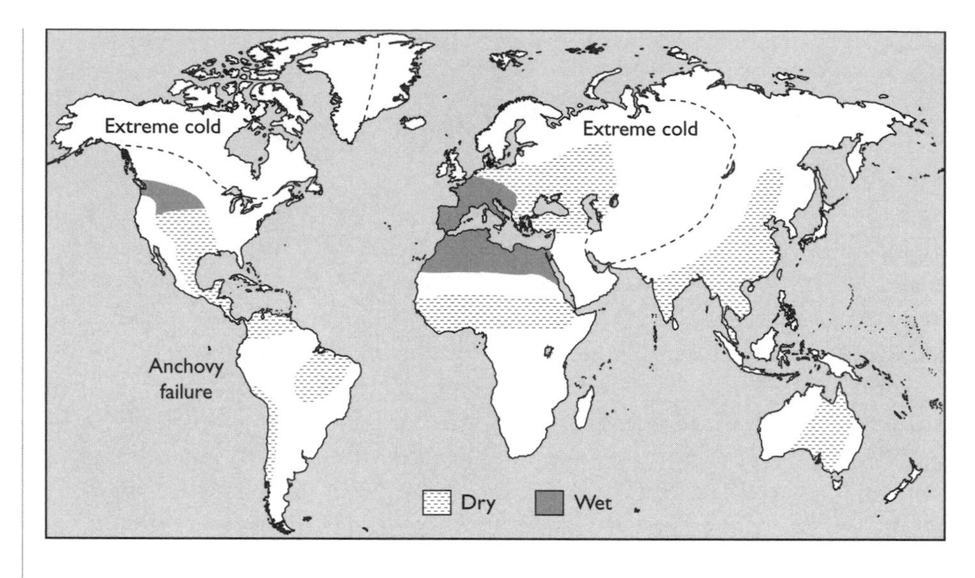

Figure 98 Strange weather in 1972 caused by El Niño. Stippled areas were affected by extreme drought. Darkened areas were affected by unusually wet weather.

unusual weather was blamed on an El Niño event, which is a warming of the eastern Pacific Ocean that occurs every three to eight years (Fig. 98). In terms of economic costs and human misery, the phenomenal storms could go down in the record books as possibly the worst weather event of the 20th century.

Another strong El Niño event appears to have been responsible for the Midwest floods in the United States during the spring and summer of 1993. The jet stream, which normally sweeps far south in winter and far north in summer, became stationary over the upper Midwest, where it steered strong weather systems into the region. Major rivers such as the Mississippi and Missouri overran their levees and drowned adjacent floodplains. Tens of thousands of people were left homeless, and tens of millions of acres of cropland were destroyed.

The disaster, designated as a 500-year flood, was the costliest in the nation's history. It amounted to $15 to $20 billion in damages and resulted in 48 lives lost. The flooding is considered a man-made disaster because levees constructed to protect property severely restrict river flow during flood stage, and reservoirs built to contain normal floods tend to overflow during massive flooding. Furthermore, upstream floodplains and wetlands act as a sponge to soak up excess floodwaters. Levees restrict this function, causing serious flooding downstream.

Another 500-year flood occurred in early April 1997 on the Red River of the North bordering North Dakota and Minnesota. Record snowfall and rapid spring melting and ice jams combined to drive the river up to 50 feet or more above flood stage, where levees restricted its flow. As the river approached the Canadian border, it overflowed its channel and stretched 40 miles across one of the flattest valleys on Earth. The flood caused more than a billion dollars in property damages and forced some 100,000 people from their homes.

FLOOD TYPES

Flash floods are local floods of great volume and short duration. They are the most intense form of flooding and generally result from torrential rains or cloudbursts associated with severe thunderstorms over a relatively small drainage area. Flash floods also occur after dam breaks or by the sudden breakup of ice jams, causing the release of large volumes of flow in a short duration. An unusual type of flash flood resulted from the 1980 Mount St. Helens eruption. It produced major mudflows and flooding from melted glaciers and snow on the volcano's flanks, sending a torrent of mud and water laden with tree trunks into nearby rivers.

Flash floods can strike in almost any part of the nation. However, they occur especially often in the mountainous areas and desert regions of the American West. Floodwaters rapidly flowing out of dry mountain regions carry a heavy sediment load, including blocks the size of automobiles. When the stream reaches the adjacent desert, its water rapidly percolates into the desert floor, bringing it to an abrupt halt. Sometimes huge monoliths hauled out of the mountains dot the desert landscape as tributes to the tremendous power of water in motion.

Flash floods are particularly dangerous in areas where the terrain is steep, surface runoff rates are high, streams flow in narrow canyons, and severe thunderstorms are prevalent. Flash floods from violent thunderstorms produce flooding on widely dispersed streams, resulting in high flood waves. The discharges quickly reach a maximum and diminish almost as rapidly. Floodwaters frequently contain large quantities of sediment and debris collected as the river sweeps clean the stream channel.

Often, cities with well-designed drainage systems that can handle normal high-water levels are totally swamped by a flash flood. The water level rises too rapidly for the drains to remove the excess, and they become overloaded with water that overflows into the streets. Runoff from intense rainfalls can result in high flood waves that destroy roads, bridges, homes, buildings, and other community developments.

Riverine floods (Fig. 99) are caused by heavy precipitation over large areas, by the melting of a winter's accumulation of snow, or both. They differ from flash floods in extent and duration. Riverine floods occur in river systems whose tributaries drain large geographic areas and encompass many independent river basins. Floods on large river systems might continue for a

Figure 99 *Extensive flooding on the Feather River, Sutter County, California.*

(Photo by W. H Hoffman, courtesy USGS)

few hours to several days. The floods are influenced by variations in the intensity and the amount and distribution of precipitation. Other factors that directly affect flood runoff are the condition of the ground, the amount of soil moisture, the vegetative cover, and the amount of urbanization, specifically impervious pavement, which exacerbates runoff.

Upstream floods in the upper parts of a drainage system are produced by intense rainfall of short duration over a relatively small area. They generally do not cause floods in the larger rivers they join downstream because of their greater stream capacity. Conversely, downstream floods cover a wide area and are usually caused by storms of long duration that saturate the soil, resulting in increased runoff. The contribution of additional runoff from many tributary basins can produce flooding downstream, recognized by the migration of an ever-increasing flood wave with a large rise and fall in discharge.

The movement of floodwaters is controlled by the size of the river and the timing of flood waves from tributaries emptying into the main channel. As a flood moves down a river system, temporary storage in the channel reduces the flood peak. As tributaries enter the main channel, the river enlarges downstream. Since tributaries are different sizes and not spaced uniformly, their flood peaks reach the main channel at different times, thereby smoothing out the peaks as the flood wave moves downstream.

Tidal floods are overflows on coastal areas bordering the ocean, an estuary, or a large lake. The coastal lands, including bars, spits, and deltas, are affected by coastal currents and offer the same protection from the sea that floodplains do from rivers. Coastal flooding is primarily a result of high tides, waves from high winds, storm surges (Fig. 100), tsunamis, or any combination of these. Tidal floods are also caused by waves generated by hurricane winds along with flood runoff resulting from heavy rains that accompany the storms. The flooding can extend over large distances along a coastline.

The duration is usually short and depends on the elevation of the tide, which usually rises and falls twice daily. If the tide is in, other forces that produce high waves can raise the maximum level of the prevailing high tide. The most severe tidal floods are caused by tidal waves generated by high winds superimposed onto regular tides. Hurricanes are the primary sources of extreme winds. Each year, several of these storms enter the American mainland, causing a tremendous amount of damage and flooding as well as severe beach erosion that continues to move the coastline landward.

Figure 100 *Hurricane storm surge damage to homes at Virginia Beach, Virginia, in March 1962.*

(Courtesy NOAA)

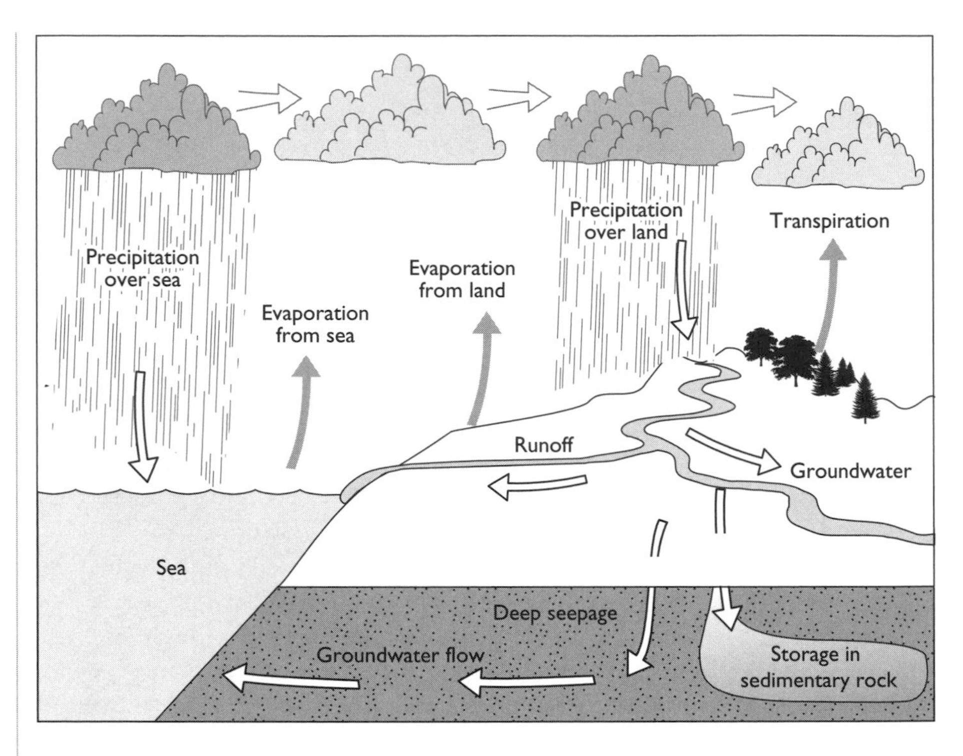

Figure 101 *The hydrologic cycle involves the flow of water from the ocean onto the land and back into the sea.*

THE HYDROLOGIC CYCLE

The movement of water from the ocean, over the land, and back to the sea again is known as the hydrologic cycle (Fig. 101). It is one of nature's most important cycles because without the flow of water over the Earth, life as we know it could not exist. The oceans cover more than 70 percent of the planet's surface to an average depth of over 2 miles. The total volume of seawater is nearly a quarter billion cubic miles.

Water travels from the ocean to the atmosphere, crosses the land, and empties into the sea. This entire journey takes, on average, about 10 days. The journey is only a few hours long in the tropical coastal areas and up to 10,000 years in the polar regions. Snow falling onto the polar ice sheets reenters the sea as icebergs when glaciers plunge into the ocean. The quickest route water takes to the sea is by runoff from streams and rivers. This is the most apparent as well as the most important part of the hydrologic cycle. Rivers provide waterways for commerce and water for hydroelectric power, municipal water supplies, recreation, and irrigation (Fig. 102). Consequently, many of the world's major cities are situated near waterways.

Every day, a trillion tons of water rain down onto the surface, mostly falling back into the sea. The total amount of precipitation the land receives is

Figure 102 *River water irrigation near Grand Junction, Colorado.*

about 25,000 cubic miles annually. Some 10,000 cubic miles are surplus water lost by floods, held by soils, or contained by wetlands. About a third of the total is base flow, which is the stable runoff of all the world's rivers and streams. The rest is groundwater flow. Some 15,000 cubic miles of water evaporate from lakes, rivers, aquifers, soils, and plants per year.

Much of the water the continents receive is lost by floods, which are important for the distribution of soils over the land. During a flood, a river might change course many times as it meanders along on its journey to the sea. Surface runoff supplies minerals and nutrients to the ocean and cleanses the land. The importance of water to life is obvious, yet it is too often overlooked. Much of the surface and subsurface water has become polluted by human activities. Additionally, the accumulation of toxic substances in the ocean from polluted runoff could cause irreparable damage to marine life.

SURFACE RUNOFF

Rivers are continuously evolving entities that adapt to environmental pressures. Running water is responsible for altering the landscape more than any other natural process. Rivers erode valleys and provide a system of drainages delicately balanced with the climate, topography, and lithology. Rivers transport sediments delivered to the channel by tributaries and by slope erosion on the valley sides. The sediment carried by rivers is temporarily stored by deposition in the channel and on the adjacent floodplain. Streams, heavily laden with sediments, overflow their beds, forcing them to detour as they meander toward the sea.

The world's rivers carve out a rugged landscape and are the primary agents for transporting the products of erosion. The river content consists of suspended load, bed load, and dissolved load. The suspended load is fine material that slowly settles out and is therefore carried long distances. The total amount of sediment in suspension increases downstream as more tributaries enter the river. The suspended load is nearly two-thirds of the total river content and amounts to about 25 billion tons per year.

The bed load is material such as pebbles and boulders that travel by rolling and sliding along the river bottom during high flow or flood and is a quarter or less of the total river content. The dissolved load derived from chemical weathering and from solution by the river itself makes up about 10 percent of the total river content. When rivers reach the ocean, their velocity falls off sharply and their sediment load drops out of suspension, continually building the continental margins outward.

River erosion is caused by abrasion and solution. Abrasion occurs when the transported material scours the sides and bottom of the channel. A com-

Figure 103 *Potholes on the surface of a large rock mass above the mouth of Blue Creek, Tuscaloosa County, Alabama.*

(Photo by C. Butts, courtesy USGS)

mon type of abrasion forms kettle-shaped depressions called potholes (Fig. 103) worn in riverbeds by rocks spinning around inside by turbulent eddy currents. Many of the largest potholes formed during the melting of glaciers after the Ice Age. The impact and drag of the water itself also erode and transport material. Most of the dissolved matter in a stream originates from groundwater draining from a breached water table. Materials such as limestone dissolve in slightly acidic river water. Limestone also acts as a buffer to maintain acidity levels within tolerable limits for aquatic life.

Erosion tends to deepen, lengthen, and widen river valleys. At the head of a stream, where the slope is steep and water flows quickly, downcutting lengthens the valley by a process called headward erosion, which is mainly how streams cut into the landscape. Farther downstream, both the velocity and discharge increase, while the sediment size and the number of banks decrease. This allows the river to transport a larger load with lesser slope. Bends in the stream tend to slow the river flow by lowering the gradient.

The most effective means by which erosion widens a stream valley are creep, landsliding, and lateral cutting. These processes are most pronounced on the outsides of irregular curves, where the valley side might be undercut by flowing water. Therefore, streams with migrating curves tend to widen their valleys. Many streams have distinctive symmetrical curves called meanders

Figure 104 *Classic meanders of Crooked Creek, Mono County, California.*

(Photo by W. T. Lee, courtesy USGS)

(Fig. 104) that distribute the river's energy uniformly. Intersecting meanders form isolated cutoff sections of the river called oxbow lakes.

Erosion rates depend on the rainfall, evaporation, and vegetative cover in the drainage basin. The ability of a river to erode and transport material depends largely on the velocity, the water flow, the stream gradient, and the shape and roughness of the channel. The average rate of erosion in the United States is about 2.5 inches per 1,000 years. The Columbia River basin has the lowest erosion rate at 1.5 inches per 1,000 years, and the Colorado River basin has the highest at 6.5 inches per 1,000 years.

DRAINAGE BASINS

A drainage basin comprises the entire area from which a stream and its tributaries receive water. For example, the Mississippi River and its tributaries drain an enormous section of the central United States from the Rockies to the Appalachians. Furthermore, all tributaries emptying into the Mississippi have their own drainage areas, becoming parts of a larger basin.

Every year, about 25 billion tons of sediment are carried by stream runoff into the ocean, where it settles onto the continental shelf. The sediment is produced when rocks weather by the action of rain, wind, and ice. Loose sediment grains are then carried downstream to the ocean. Rivers such as the Amazon and Mississippi transport enormous quantities of sediment derived from the interiors of their respective continents.

Each year, the Mississippi River dumps over a quarter billion tons of sediment into the Gulf of Mexico, widening the Mississippi Delta and slowly building up Louisiana and nearby states (Fig. 105). The Gulf coastal states from East Texas to the Florida panhandle were built with sediments eroded from the interior of the continent and brought down by the Mississippi and other rivers. The Amazon, the world's largest river, is forced to carry heavier sediment loads due to large-scale deforestation and severe soil erosion at its headwaters. India's Ganges and Brahmaputra Rivers carry about 40 percent of the world's total amount of sediment discharged into the ocean as erosion gradually wears down the Himalaya Mountains.

Individual streams and their valleys join into networks that display various types of drainage patterns, depending on the terrain. In areas of exposed bedrock, drainage patterns depend on the lithology of the underlying rocks, the attitude of rock units, and the arrangement and spacing of planes of weak-

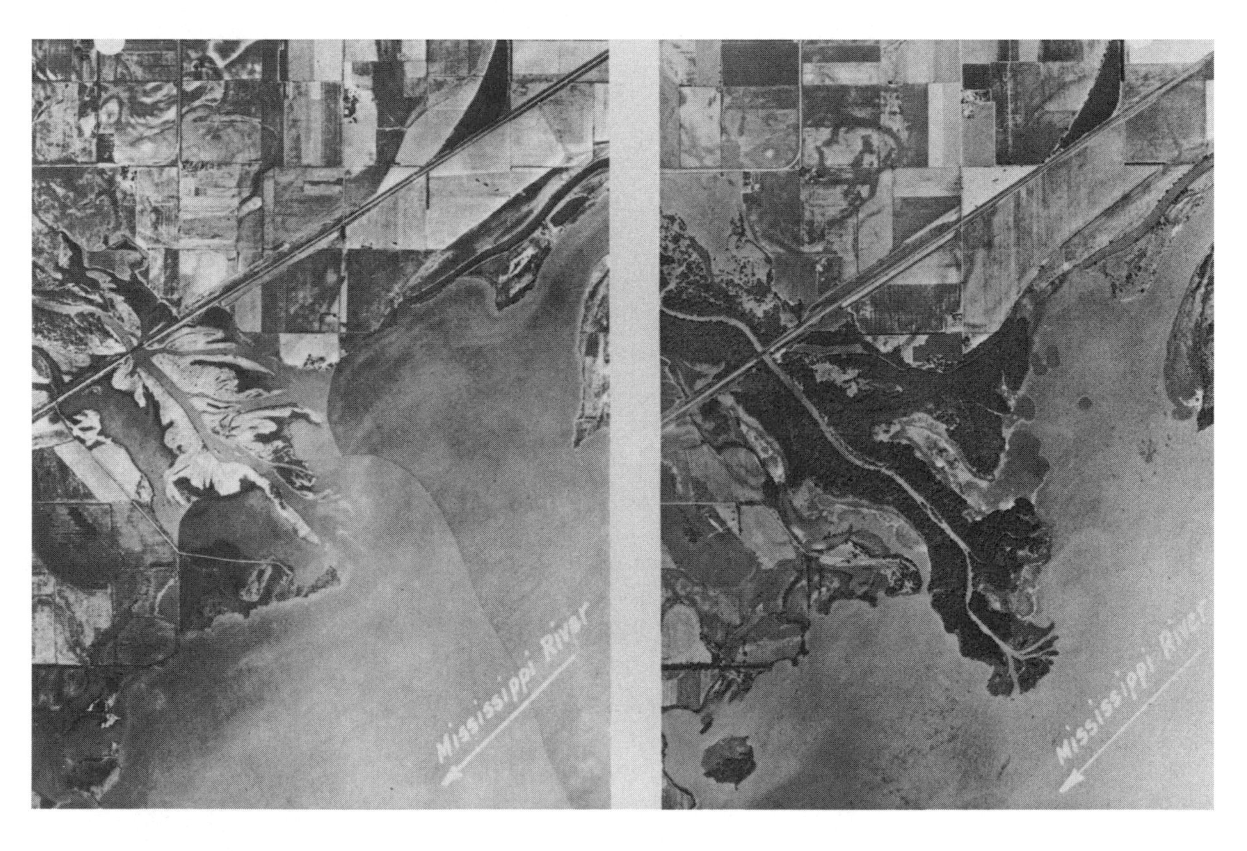

Figure 105 *Sediment deposition in the Mississippi River Delta: 1930 conditions* (left), *1956 conditions* (right).
(Photo by H. P. Guy, courtesy USGS)

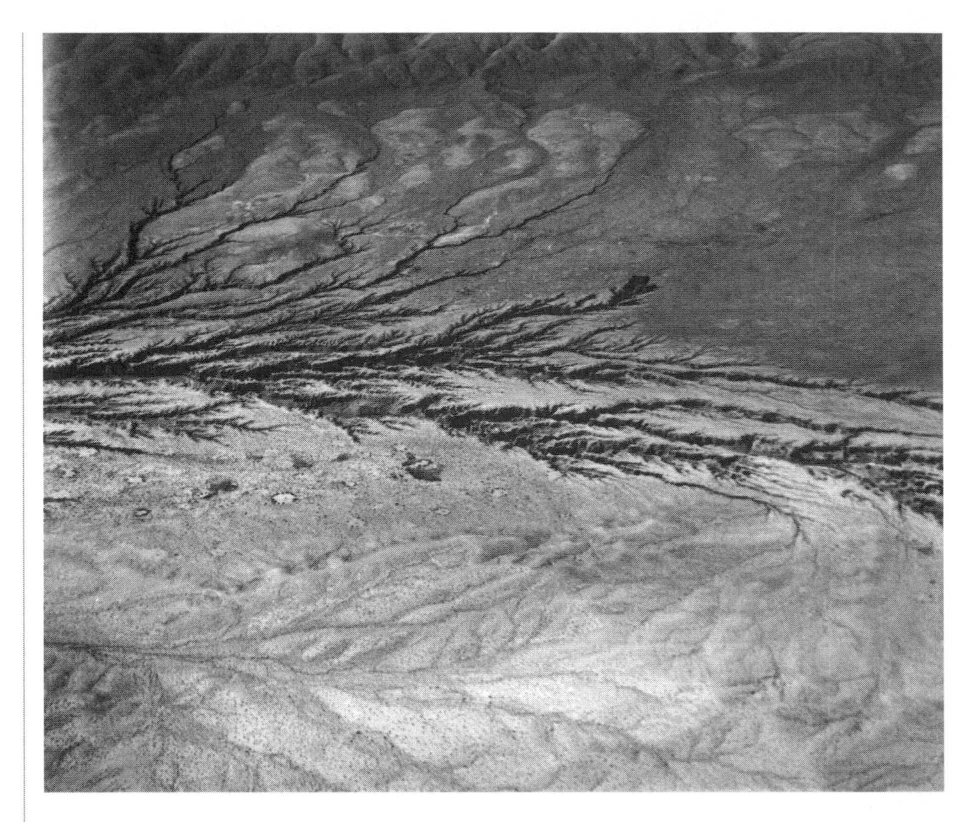

Figure 106 *A dendritic drainage pattern near Green River, Utah.*

(Photo by J. R. Balsley, courtesy USGS)

ness encountered by runoff. If the terrain has a uniform composition and does not determine the direction of valley growth, the drainage pattern is dendritic (Fig. 106), resembling the branches of a tree. Granitic and horizontally bedded sedimentary rocks generally yield this type of stream pattern.

A trellis drainage pattern displays rectangular shapes that reflect differences in the bedrock's resistance to erosion. The major tributaries of the master stream follow parallel zones of least resistance over underlying folded rocks. Rectangular drainage patterns also occur when fractures crisscross the bedrock, forming zones of weakness that are particularly susceptible to erosion. If streams radiate outward in all directions from a topographical high such as a volcano or dome, they produce a radial stream drainage pattern.

Drainage patterns are influenced by topographical relief and rock type. They provide important clues about the geologic structure of an area. In addition, the color and texture of the structure impart information about the rock formations that compose it. Surface expressions such as domes, anticlines, synclines, and folds bear clues about the subsurface structure. Various types of drainage patterns imply variations in the surface lithology or rock type. The drainage pattern density is another indicator of the lithology.

Variations in the drainage density also correspond to changes in the coarseness of the alluvium.

River deposits called alluvium accumulate due to a decline in stream gradient or inclination, a reduction in stream flow, or a decrease in stream volume, with the heaviest material settling out first. Changes in the river environment occur during entrance into standing water, encounters with obstacles, evaporation, and freezing. River deposition is divided into deposits in bodies of water, alluvial fans, and deposits within the stream valley itself. A medium-sized river will take about a million years to move its sandy deposits 100 miles downstream. Along the way, the grains of sand are polished to a high gloss.

Sedimentary rocks deposited within streambeds occur relatively rarely because rivers deliver most of their sediment load to lakes or the sea. River deltas (Fig. 107) develop where rivers enter larger rivers or standing water. The initial velocity of the river slows so abruptly when entering a body of water that its bed load immediately drops out of suspension. Much of the river's load is also reworked by offshore currents, creating marine or lake deposits.

Alluvial fans generally found in arid regions are similar to river deltas. They form where streams flow out of mountains onto broad valleys, where the ground abruptly flattens. This causes the stream to slow and deposit its sediment load in a fan-shaped body. As the stream constantly shifts its position, the alluvial fan grows steeper, thicker, and coarser, developing a characteristic cone shape.

Channel fill is alluvium laid down in the channel of a stream. Accumulations of fill assume many shapes generally known as sandbars. They collect

Figure 107 *The delta of Chelan River entering the Columbia River, Chelan Ferry, Washington, September 18, 1900.*
(Photo by B. Willis, courtesy USGS)

along the edges of a stream, especially on the insides of bends, accumulate around obstructions, and pile up into submerged shoals and low islands. These deposits are not permanent features but are destroyed, redeposited, or shift positions as conditions on the river change.

Natural levees build up on the banks of a river during a flood. When floodwaters overtop a river channel and flow onto the adjacent floodplain, the velocity quickly diminishes depositing sediments near riverbanks. The riverbanks provide a variety of environments for plant life, which helps stabilize them. The inflow of nutrients and sediments; changing water levels over the seasons, which can create different biologic niches; and waterborne dispersal of seeds contribute to a rich diversity of species occupying riverbanks.

Levees help keep the river within its banks during normal flow. However, at flood stage, the valley floor is often lower than the river level, inundating the land when floodwaters crest over the levee top. During the 1993 Midwest floods, perhaps the worst of the 20th century, levee breaks deposited several feet of sand onto farmlands along the swelling rivers. Artificial levees continue to break during major floods, compounding the amount of death and destruction.

When a river excavates part of its floodplain due to changes in flow, it leaves a terrace standing above the river's new level. Terraces first develop in the lower reaches of a stream and then extend upstream, cutting into sediments laid down earlier in either direction. Terraces also form by lateral cutting of bedrock by a river. One of the most common causes of terrace building is associated with glacial meltwater during the end of the last Ice Age. Melting glaciers produced more debris than rivers could handle. The excess sediment was deposited into the river valleys. As conditions returned to normal, terraces formed when rivers downcut into these deposits.

An unusual type of river flow, called a braided stream, forms when the bed load is too large and coarse for the slope and the amount of discharge. The banks easily erode, which chokes the channel with sediment, causing the stream to divide and rejoin repeatedly. The stream deposits the coarser part of its abundant load to attain a steep enough slope to transport the remaining load, which forces the stream to broaden and erode its banks. Alluvium rapidly deposits in constantly shifting positions, forcing the stream to split into interlacing channels that continuously separate and reunite.

A river widens its valley as it flows along a leveling grade and is no longer rapidly downcutting, referred to as the mature stage. This condition occurs mostly near the mouth of a river, where wide floodplains exist. Meanders are common features of wide valleys, especially in areas with uniform banks composed of easily erodible sediment. The valley might widen by flooding, weathering, and mass wasting. Many river valleys were also widened by glaciers during the Ice Age, converting V-shaped valleys into U-shaped ones (Fig. 108).

Figure 108 *A U-shaped glaciated valley at Red Mountain Pass south of Ouray, Colorado.*

(Photo by L. C. Huff, courtesy USGS)

Rivers clogged with sediment fill their channels, spill over onto the adjacent plain, and carve out a new river course. In the process, they meander downstream, forming thick sediment deposits in broad floodplains that can fill an entire valley. As a stream meanders across a floodplain, the greatest erosion takes place on the outside of the bends. This results in a steep cut bank in the channel. On the inside of the bends, however, the water slows and deposits its suspended sediments. During a flood, a winding river often takes a shortcut across a low-lying area separating two bends, temporarily straightening the river until it further fills its channel with sediment, causing it to meander once again. Meanwhile, the cutoff sections of the original river bends become oxbow lakes.

A river also might capture a nearby stream, known as "piracy," creating a larger expanse of flowing water. The river grows at the expense of other streams and becomes dominant because it contains more water, erodes softer rocks, or descends a steeper slope. The river therefore has a faster headward erosion that undercuts the divide separating it from another stream and captures its water.

FLOOD-PRONE AREAS

Lateral migration of bends of rivers and overbank flow combine to produce the floodplain, which is periodically inundated by water and sediment during floods. Channel discharge where water overflows the riverbank is called the flood stage. This high-water condition often causes property damage on the floodplain. Unfortunately, only when a flood occurs do people finally recognize the importance of good floodplain management so as to reduce the severity of floods.

Floods are typically natural disasters because people continue to build on floodplains without recognizing the flood potential. Floodplain zoning laws and flood control projects are based on statistical analyses of relatively short-term historical records of large floods, sometimes referred to as 100- or 50-year floods. This makes assessing the risk of large floods very difficult. Moreover, due to variability in the weather, two or more record-breaking floods can occur in consecutive years.

The purpose of floodplains is to carry off excess water during floods. Failure to recognize this function has led to haphazard development in these areas with a consequent increase in flood dangers. Floodplains provide level ground, fertile soils, ease of access, and available water supplies. However, because of economic pressures, these areas are being developed without full consideration of the flood risk. As a consequence, the United States Federal Government has assumed much of the responsibility for providing flood relief.

In spite of flood protection programs, the average annual flood hazard has been on the rise because people have been moving into flood-prone areas faster than the construction of flood protection projects. Therefore, the increased losses are not necessarily the result of larger floods but of greater encroachment onto floodplains. As the population rapidly increases, more pressure is applied to develop flood-prone areas without taking proper precautions. Too often, people are uninformed about the flood risk when they build in flood-prone areas. When the inevitable flood strikes, they turn to the government to pay the cost of rebuilding.

About 3.5 million miles of rivers and streams flow in the United States, and about 6 percent of the land is prone to flooding (Fig. 109). A large percentage of the nation's population and property is concentrated in these flood-prone areas. More than 20,000 communities have flood problems. Of these, about 6,000 have populations of more than 2,500. Due to high population growth, modern floods are becoming more hazardous. For example, both the 1973 and 1993 Mississippi River floods, the 1978 Pearl River flood in Louisiana and Mississippi, and the 1997 Red River of the North flood were four of the costliest floods in American history.

Floods not only threaten lives but also cause much suffering, damage property, destroy crops, and halt commerce. The average annual flood loss in the United States has increased from less than $100,000 at the turn of the 20th century to some $4 billion today. Presently, floods take some 100 lives in the United States annually. Many lives could have been saved if proper precautions were taken in flood-prone areas.

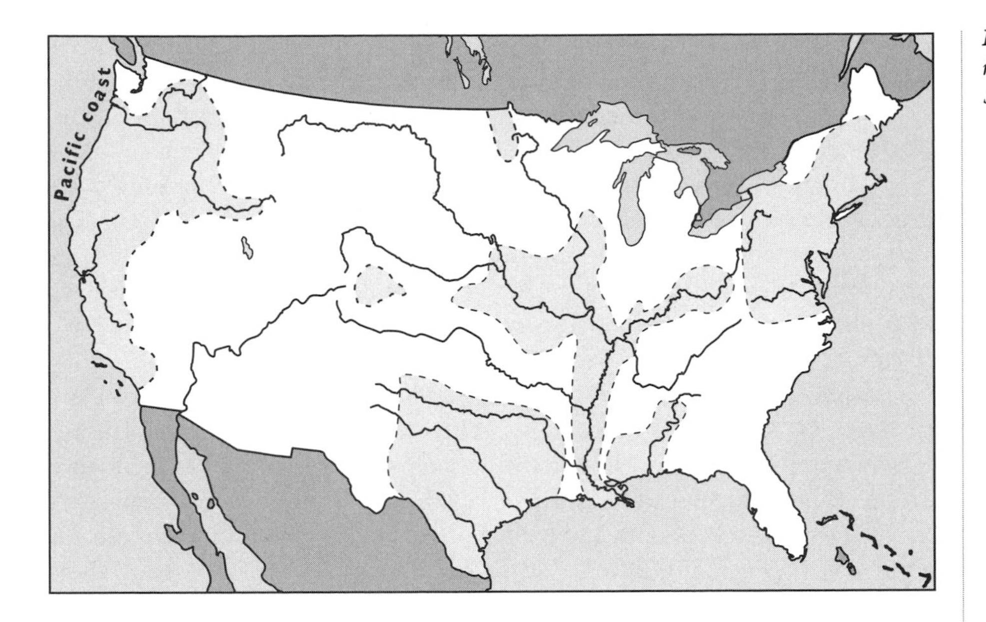

Figure 109 *Flood-prone regions of the United States.*

HYDROLOGIC MAPPING

Hydrologic mapping using satellite imagery provides useful data about snow cover, sea ice extent, river flow, and flood inundation. The 1973 and 1993 Mississippi River floods, the 1978 Kentucky River flood, and the 1978 and 1997 Red River of the North floods resulted in near record-breaking water levels identified on satellite imagery. Regional snow cover maps are important for predicting the amount of runoff during spring thaw. The unusually stormy winter of 1982–1983 produced a larger-than-normal snowpack in the Rocky Mountains of Colorado. During the spring thaw, the Colorado River reached record flood stage. A repeat performance occurred in 1993.

Images obtained from weather satellites are used to monitor river basin snow cover in the United States and Canada. Several government agencies and private concerns, such as utility companies, use the river basin snow maps to predict water availability. The snow data also aid in dam and reservoir operations as well as help to calibrate runoff models. These models are designed to simulate and forecast daily streamflow in basins where snowmelt is a major contributor to runoff. This is particularly important in the West for preparing seasonal water supply forecasts.

Several methods are used to analyze satellite data for mapping snow cover. The simplest uses an optical transfer device to magnify and rectify the satellite imagery so it overlies a standard hydrologic basin map. The snow line is then manually transferred from the image to the map. Another method uses a computer to display the image data on a video monitor, and the snow line is electronically traced onto the image. A third method uses a computer to determine the snow cover pixel by pixel (tiny picture elements) by analyzing the terrain type and solar incidence angles, which control the amount of sunlight reflected back to space.

The snow maps display the extent of continental snow cover but do not indicate the snow depth, which must be obtained manually in the field. The snow cover charts are digitized and stored on computer disks, from which are created monthly anomaly, frequency, and climatological snow cover maps. In addition, continental or regional snow cover can be calculated over long periods for North American winter snow cover.

Satellite data are useful for detecting and locating ice cover and ice dams on rivers, especially northern streams, where ice is particularly troublesome. Observation of river ice is important because it creates problems for hydroelectric dams, bridges, and maritime navigation. The ice becomes particularly hazardous when it breaks up and forms a dam, posing a flood threat to nearby communities (Fig. 110). Often, the ice persists because of river dams, sharp bends in the river course, or branching of the main channel by islands.

The data are also used for monitoring flash floods from large storm systems. Satellite-derived precipitation estimates and trends aid meteorolo-

Figure 110 *An ice jam in the Passumpsic River, causing flooding at St. Johnsbury Center, Vermont.*

(Courtesy USDA Soil Conservation Service)

gists and hydrologists in evaluating heavy precipitation events and in providing timely warnings to affected areas. Flood damage in the United States often exceeds a billion dollars annually despite the construction of flood prevention projects to help save lives and reduce property losses. To minimize flood-related hazards, engineers and governmental officials need accurate information on the location of flood hazard areas and assessments of areas of inundation when floods occur. Computer models are used to provide quick approximations of the total extent of a flood for disaster and relief planning.

FLOOD CONTROL

The dangers of flooding can be vastly alleviated by taking certain precautions that ultimately save lives and property. The factors that control damage arising from floods include land use on the floodplain; the depth and velocity of the water and the frequency of flooding; the rate of rise and duration of flooding; the time of year; the amount of sediment load deposited; and the effectiveness of storm forecasting, flood warning, and emergency services.

Direct flood effects include injury, loss of life, and damage to buildings and other structures caused by swift currents, debris, and sediment. In addition, sediment erosion and deposition on the landscape might involve a considerable loss of soil and vegetation. Indirect flood effects include short-term pollution of rivers, the disruption of food supplies, the spread of disease, and the displacement of people who have lost their homes. In addition, floods might cause fires due to short circuits in electric lines or breaks in gas mains.

Flood prevention involves engineering structures such as levees and flood walls that serve as barriers against high water (Fig. 111), building reservoirs to store excess water for later release at safe rates, increasing the channel size to move water quickly off the land, and diverting channels to route floodwaters around areas that require protection. The best method of minimizing flood damage in urban areas is floodplain regulation along with barriers, reservoirs, and channel improvements in flood-prone areas already developed to protect lives and property. In response to the damage caused by the 1993 Midwest flood, the federal government paid homeowners to move their houses to higher ground.

Structures such as levees tend to aggravate floods by forcing rivers into narrow channels instead of allowing the floodwaters to drain naturally onto floodplains, where the power of the floods is dissipated. The effects of flooding are significantly reduced by developing a flood control system based on the addition of natural wetlands that absorb excess floodwaters as well as less river engineering and levee construction. Wetlands can have a marked impact on the containment of most floods but are less useful during great floods, especially if heavy rains have already fully saturated the ground.

Municipalities should also cease development of areas subject to flooding that require new barriers. The most practical solution is a combination of floodplain regulations and barriers that result in less physical modification of the river system. Therefore, reasonable floodplain zoning might result in lesser use of flood prevention methods than if no floodplain regulations exist.

The United States has spent billions of dollars on flood protection projects. Most of these are artificial reservoirs (Fig. 112) that even out the flow rates of rivers and have a storage capacity that can absorb increased flow during floods. The dams also generate hydroelectric power. In addition, their reservoirs provide river navigation, irrigation, municipal water supplies, fisheries, and recreation. However, without proper soil conservation measures in the catchment areas, the accumulation of silt by erosion can severely limit the life expectancy of a reservoir.

Floodplain regulations are designed to obtain the most beneficial use of floodplains while minimizing flood damage and the cost of flood protection. The first step in floodplain regulations is flood hazard mapping, which provides floodplain information for land use planning. The maps delineate past floods and help derive regulations for floodplain development. These controls

Figure 111 *Workers repair an overtopped back levee near New Orleans, Louisiana.*

(Courtesy Army Corps Engineers)

Figure 112 *Hoover Dam and Lake Mead on the border of Nevada and Arizona.*

(Photo courtesy USGS)

are a compromise between the indiscriminate use of floodplains that results in the destruction of property and loss of life and the complete abandonment of floodplains, thereby relinquishing a valuable natural resource. Only by recognizing the dangers of flooding and preparing for the worst can people safely utilize what nature has reserved for excess water during a flood.

After discussing the effects of floods and their danger to society, the next chapter takes a look at a drier environment—where dust storms are prevalent in the world's deserts along with desert erosion, sandstorms, sand dunes, and dust bowls.

7

DUST STORMS

SAND IN MOTION

This chapter examines the geologic hazards in the desert regions of the world. Dust storms are awesome meteorologic events that play a significant role in people's lives in many parts of the world. They can directly threaten life; both people and animals have died of suffocation during severe dust storms. Another direct threat dust storms pose is soil erosion. Topsoil is literally tossed to the wind and carried long distances away.

The Dust Bowl years of the 1930s were the nation's worst ecological disaster (Fig. 113). Tremendous quantities of topsoil were airlifted out of the American Great Plains and deposited elsewhere, often burying areas under thick layers of sediment. Massive dust storms raced across the prairie, carrying more than 150,000 tons of sediment per square mile. Since then, agricultural practices have decreased this hazard in the United States as well as in other parts of the world. Unfortunately, many regions are still at risk from dust storms and accompanied soil erosion, which is seriously undermining efforts of populations to feed themselves.

Figure 113 *Buried machinery on a farm in Gregory County, South Dakota, during the 1930s Dust Bowl.*

(Courtesy USDA Soil Conservation Service)

DESERT REGIONS

Deserts are among the most dynamic landscapes, constantly changing by drifting sands and other geologic processes unique to the desert environment. Gigantic dust storms and sandstorms prevalent in desert regions play a major role in shaping the arid terrain. Powerful sandstorms clog the skies with thousands of tons of sediment. Roving sand dunes driven across the desert by strong winds engulf everything in their paths. Coastal deserts are unique because they are areas where the seas meet the desert sands, as exemplified by the Namib Desert (Fig. 114) along the coast of Namibia, Africa, perhaps the world's largest coastal desert.

In the last Ice Age, lowered precipitation levels caused desert regions to expand in many parts of the world. Desert winds blew much stronger than they do today, promoting gigantic dust storms that blocked out sunlight, furthering cool conditions. When the great ice sheets began to retreat to the poles, tropical regions of Africa and Arabia began to dry out during a period of rapid warming, resulting in the expansion of deserts between 14,000 and 12,500 years ago.

During an unusual wet spell from 12,000 to 6,000 years ago, many of today's African deserts were lush with vegetation and contained several large lakes. Lake Chad on the southern border of the Sahara Desert swelled several times its present size. The Mojave and nearby deserts of the American Southwest received sufficient rainfall to sustain woodlands. Utah's Great Salt Lake

Figure 114 *Linear dunes in the northern part of the Namib Desert, Namibia.*
(Photo by E. D. McKee, courtesy USGS)

expanded well beyond its present shores to occupy the adjacent salt flats. The Climatic Optimum, which began about 6,000 years ago, was a period of unusually warm, wet conditions that lasted 2,000 years. Then around 4,000 years ago, temperatures dropped significantly and the world's climate became dryer, spawning the return of the deserts.

The study of fossilized pollen has shown that grasses and shrubbery covered what is now the Sahara Desert until some unknown environmental catastrophe dried up all the water, leaving behind nothing but sand. A relatively mild arid episode between 7,000 and 6,000 years ago was followed by a severe 400-year-long drought starting 4,000 years ago. Apparently, the monsoon storms that provided water to the Sahara grew weaker, killing off native plants. The reduction in vegetation further reduced rainfall, producing a vicious cycle of desertification. The drought caused by the vegetation feedback mechanism consequently wiped out almost all plant and animal life in the desert. Such a disaster might have driven entire civilizations out of the desert, causing them to found new societies on the banks of the Nile, Tigris, and Euphrates Rivers.

A number of clues give researchers information about the climates of the past. Tree rings (Fig. 115) provide an ideal indicator of past climates. Generally, the wider the rings, the more favorable the climate was. During a drought or a cold climate, tree rings are usually narrower due to poor growing condi-

Figure 115 *A tree sample being prepared for annual growth ring studies.*

(Photo by L. E. Jackson Jr., courtesy USGS)

tions. By analyzing tree rings of the bristlecone pine, one of the longest-living plants, scientists have established a drought index for the western United States dating to the year 1600. By measuring tree rings of ancient, well-preserved trees, investigators can delve into the climate history extending back more than 7,000 years.

Deserts are more than just barren landscapes mostly devoid of vegetation and animal life. They are constantly changing by sand in motion. Sometimes sand dunes cover over human settlements and other man-made features, often causing considerable damage. Sandstorms are particularly hazardous as thousands of tons of sediment clog the skies and land in places where it is not wanted.

Deserts are the hottest and driest regions and among the most desolate environments on Earth. About a third of the land surface is desert (Fig. 116 and Table 6). Deserts generally receive less than 10 inches of average annual rainfall, and evaporation usually exceeds precipitation throughout the year. Because of these stringent conditions, desert areas cannot support significant human populations without artificial water supplies. Much of the world's desert wastelands receive only minor quantities of rain during certain seasons. Some regions, such as Egypt's Western Desert, have gone essentially without rain for many years.

When the rains finally do arrive, they are often violent and local, causing severe flash floods. A typical desert rain falls as a short torrential downpour that floods several square miles. Water levels in dry wadis or washes rise rapidly and fall almost as fast, as the flood wave flows through the desert. Eventually, the floodwaters either empty into shallow lakes that later dry up

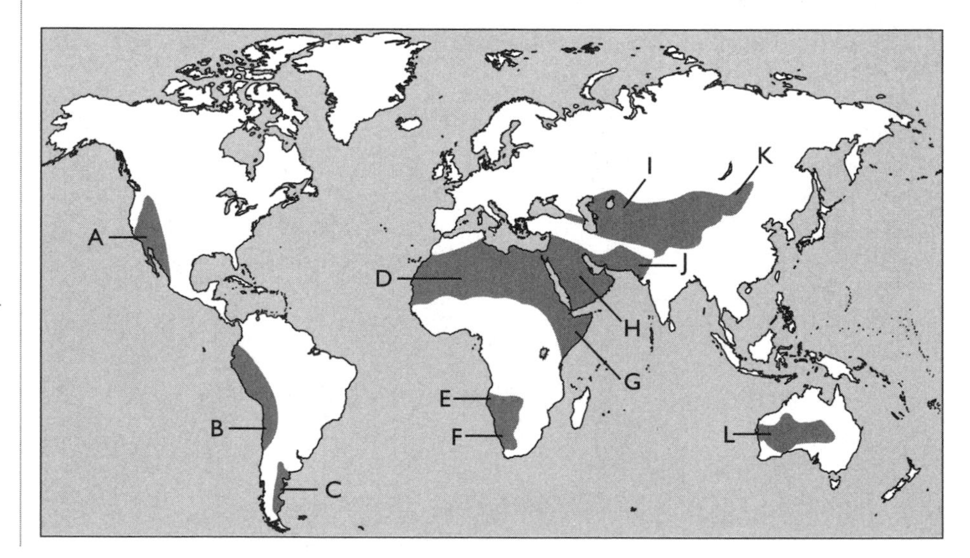

Figure 116 *The world's major deserts: A—North American, B—Peruvian-Atacama, C—Patagonian, D—Sahara, E—Namib, F—Kalahari, G—Somali, H—Arabian, I—Turkestan, J—Iranian, K—Gobi, L—Australian.*

TABLE 6 MAJOR DESERTS

Desert	Location	Type	Area (square miles × 1000)
Sahara	North Africa	Tropical	3,500
Australian	Western/interior	Tropical	1,300
Arabian	Arabian peninsula	Tropical	1,000
Turkestan	S. Central U.S.S.R.	Continental	750
North American	SW U.S./N. Mexico	Continental	500
Patagonian	Argentina	Continental	260
Thar	India/Pakistan	Tropical	230
Kalahari	S.W. Africa	Littoral	220
Gobi	Mongolia/China	Continental	200
Takliamakan	Xinjiang Uygur, China	Continental	200
Iranian	Iran/Afghanistan	Tropical	150
Atacama	Peru/Chile	Littoral	140

or soak into the dry, parched ground. Then for months or even years afterward, no rain will fall.

Most of the world's deserts lie in the subtropics between 15 and 40 degrees latitude north and south of the equator. In the Northern Hemisphere, a series of deserts stretches from the west coast of North Africa through the Arabian peninsula and Iran and on into India and China. In the Southern Hemisphere, a band of deserts runs across South Africa, central Australia, and west-central South America. After warm, moist air rises in the Tropics, where precipitation levels are high, little moisture remains for the subtropics. The dry air cools and sinks, producing zones of semipermanent high pressure. These are called blocking highs because they block advancing weather systems from entering the region, producing mostly clear skies and calm winds.

Tall mountains also tend to block weather systems by forcing rain clouds to rise, which causes them to precipitate on the windward side of the range. This produces a rain shadow zone on the lee side of the mountains, making the region rain deficient. Deserts such as those in the southwestern United States are created by such a process. Moist winds from the Pacific Ocean cool and precipitate as they rise over the Sierra Nevada and other mountain ranges in California, leaving regions to the east parched and dry. Commonly, polygonal shapes created in the desert muds (Fig. 117) form when mud contracts as it rapidly dries in the hot sun.

Figure 117 *Desiccation cracks in mudflow deposits on the Arroyo Hondo Fan, Fresno County, California.*

(Photo by W. B. Bull, courtesy USGS)

Since deserts are generally light in color, they have a high albedo (Table 7), which is the ability of objects to reflect sunlight. Desert sands absorb much heat during the day when the surface scorches at temperatures often exceeding 65 degrees Celsius. However, because the skies are generally clear at night, the thermal energy trapped in the sand quickly escapes due to the low heat capacity. This makes desert regions among the coldest environments, and even summertime temperatures at higher elevations can drop to near freezing at

TABLE 7 ALBEDO OF VARIOUS SURFACES

Surface	Percent Reflected
Clouds, stratus	
< 500 feet thick	25–63
500–1000 feet thick	45–75
1000–2000 feet thick	59–84
Average all types and thicknesses	50–55
Snow, freshly fallen	80–90
Snow, old	45–70
White sand	30–60
Light soil (or desert)	25–30
Concrete	17–27
Plowed field, moist	14–17
Crops, green	5–25
Meadows, green	5–10
Forests, green	5–10
Dark soil	5–15
Road, blacktop	5–10
Water, depending upon sun angle	5–60

night. As a result, deserts have the highest temperature extremes of any environment.

Only the hardiest species live in the deserts, including plants whose seeds can survive a 50-year drought and rodents that spend their entire lives without taking a single sip of water. They survive instead solely off the water generated by their body's metabolism. Plants and animals use a variety of adaptations to survive desert conditions. These survival techniques generally rely on the conservation of water and suspended animation during the driest part of the season.

Many plants such as the giant saguaro cactus (Fig. 118), common in the Sonora Desert of northwest Mexico and southwest United States, store water in their trunks. Other plants extract moisture such as morning dew directly from the air. During the hottest part of the day, many animals retreat to underground burrows, where the temperature difference is significant. Even the space a few feet above the ground drops several degrees. Animals perch on small bushes to take advantage of the markedly cooler air.

Figure 118 *Saguaro cactus and other vegetation on the west slope of Superstition Mountains, Pinal County, Arizona.*

(Photo by W. B. Hamilton, courtesy USGS)

During the short rainy season, aquatic species such as fish and amphibians must quickly lay their eggs before the ponds they briefly inhabit dry out. The animals then burrow into the bottom mud and lie dormant until the rains return. During the next rainy season, the animals revive, their eggs hatch, and the cycle begins anew.

Air-breathing fish are known to leave their drying ponds and walk sometimes considerable distances to new water holes. Lungfish living in African swamps that seasonally dry out hole up for long stretches until the rains return. They burrow into the moist mud, leaving an air cavity leading to the surface, and live in suspended animation, breathing with primitive lungs. Thus, they can survive out of water for several months or even a year or more if necessary. When the rainy season returns and the pond refills, the fish revive, breathing normally with gills.

The Namib Desert is inhabited by tiny fairy shrimp, whose eggs can lie dormant for as long as 20 years or more. After a rare rain shower fills the dry, shallow basins, they subsequently become teeming with life. The shrimp must quickly lay their eggs before the water evaporates in the hot desert sun, leaving the pools once again cracked and parched.

The dry valleys of Antarctica between McMurdo Sound and the Transantarctic Range are not only among the coldest places but also possibly the most impoverished deserts on Earth. They receive less than 4 inches of snowfall each year, most of which blows away by strong winds that can reach 200 miles per hour and more. Only meager signs of life exist, including blue-green algae on the bottoms of small glacier-fed lakes, soil bacteria, and a species of giant wingless fly. Antarctica hosts just two flowering plant species, which have undergone population explosions recently, possibly due to a warming climate. Delicate mosses and lichens, if disturbed, take a century to recover. The discovery of lichens in tiny pores on the undersides of rocks has fueled speculation that similar life forms might inhabit the planet Mars, whose frigid terrain has many similarities.

DESERTIFICATION

The world's deserts continue to claim more land as they encroach upon and consume neighboring semidesert grasslands. Due to natural processes and human activities, additional land is rapidly becoming desertified, amounting to as much as 15,000 square miles a year, a little less than the size of California's Mojave Desert. Much of this desertification is due to naturally increasing aridity over the past several thousand years. Throughout the world, nearly twice the area of the United States has turned to desert since the dawn of agriculture some 6,000 years ago, mostly by the abuse of the land.

In North America alone, an estimated 1.1 billion acres have been desertified. Much of the American West 150 years ago was a sea of almost uninterrupted grasslands that have since turned to desert and semiarid areas. The North American Desert is the world's fifth largest of its kind. It extends irregularly from east-central Washington to northern Mexico and from the Big Bend of the Rio Grande in west Texas to the Sierra Nevada of California. It covers some 500,000 square miles, encompassing the Great Basin region and the Sonoran and Mojave Deserts.

During last 10,000 years, the Earth's climate has been unusually mild with few large perturbations, allowing for the rise of civilization. Around 5,000 years ago, the Phoenicians migrated out of the Sahara Desert and settled along the eastern coast of the Mediterranean Sea. There they established such cities as Tyre and Sidon in what is now Lebanon. The land was mountainous and heavily forested with cedars, which became the primary source of timber for the region. When the flat plains along the coast became overpopulated, people moved to the slopes, which they cleared and cultivated, severely eroding the soil. Today, very little remains of the 1,000-square-mile forest. The bare slopes are littered with the remains of ancient terrace walls used in a futile attempt to control erosion.

In northern Syria lie the remains of several once-prosperous cities that are now dead. These ancient cities prospered by converting forests into farmland and exporting olive oil and wine. After invasion by the Persians and Arabs, followed by the destruction of agriculture, up to 6 feet of soil eroded from the slopes. Today, after 1,300 years of neglect, the once-productive land is nearly completely destroyed, a man-made desert void of soil, water, and vegetation.

The Fertile Crescent, known as the cradle of civilization, between the Tigris and Euphrates Rivers in present-day Syria and Iraq, was the breadbasket of the Middle East, feeding a population of 17 to 25 million people. Today, the region is mostly an infertile desert due to overirrigation and salt accumulation in the soil by Sumerian farmers 6,000 years ago. North Africa, now mostly desert, once had lush grass and trees on the mountains. It was the breadbasket for Rome, providing grain and meat for the Roman Empire.

About 5,000 years ago, on the plains of Mesopotamia, large irrigation projects required the hard labor of hundreds of thousands of people and a system of centralized authority to rule over them. What was once a loose-knit egalitarian society, in a mere thousand years, was transformed through agriculture into an authoritarian society, equipped with kings, captains, and slaves. Huge armies of highly organized states fought each other over the control of valuable agricultural land.

The world irrigates more than 10 percent of its cropland, requiring about 600 cubic miles of water annually. The United States irrigates nearly a quarter of its farmland (Fig. 119), tripling the amount of irrigated acreage since the Second World War. Heavy use of irrigation, which not long ago turned vast stretches of America's western desert into the world's most productive agricultural land, is now ruining hundreds of thousands of acres due to soil salinization (salt buildup). Perhaps we should take a lesson from the Sumerians.

The Sahel region of central Africa was once mostly tropical forest. It lies to the south of the Sahara Desert, extending in a 250-mile-wide band from coast to coast across Central Africa. More than 1,000 years ago, nomads of the Sahel lived by hunting and herding. They cut down and burned trees to improve grazing in the region, turning natural forests into grasslands. When colonial powers carved up Africa among themselves during the 19th century, they forced the people of the Sahel to settle down and remain in the region as farmers and herders. Because of overgrazing and soil erosion, they turned the Sahel into an extensive man-made desert.

During the droughts of the 1970s and 1980s, the worst of the 20th century, the advancing sands of the Sahara Desert overran the Sahel, steadily engulfing everything in their path at a rate of 3 miles per year. In 20 years, Central Africa lost 75 percent of its grazing land to the encroaching sands of

Figure 119 *Irrigated cotton in Imperial County, California.*

(Courtesy USDA Soil Conservation Service)

Figure 120 The soil profile, showing organic-rich topsoil and sandy, infertile subsoil below. Measurements marked in feet.

(Photo by B. C. McLean, courtesy USDA Soil Conservation Service)

the Sahara. The southern extent of the desert crept as much as 80 miles farther south. A vast belt of drought to the south spreads across the continent, parching the land and starving its inhabitants.

Desertification, which severely degrades the environment, is caused mainly by human activity and climate. Soil erosion removes millions of acres of once fertile cropland and pasture from production every year. After the land loses its topsoil from erosion, only the coarse sands of the infertile subsoil remain (Fig. 120), thus creating a desert. Desertification is a global problem. However, it is most prevalent in Central Africa, where the sands of the Sahara Desert march steadily across the Sahel region. Desertification is also self-perpetuating because the light-colored sands reflect more sunlight, creating high-pressure regions that block weather systems and lessen rainfall.

The process of desertification is exacerbated because the lack of vegetation subjects the land to increased flash floods, higher evaporation and erosion rates, and dust storms that transport the soil out of the region. The denuded land has a higher albedo, which contributes to lesser rainfall and denudes more land. This causes deserts to march across once fertile regions. Worldwide, perhaps a third or more of previously fertile land is now rendered useless by erosion and desertification. Moreover, improper irrigation methods could cause from one-half to three-quarters of all irrigated land to be destroyed by soil salinization.

Global rain forests once covered an area twice the size of Europe. Today, the forested area has been reduced by half for additional cropland. As developing nations attempt to raise their standards of living, they first clear forests and drain wetlands for agriculture. Farmers clear much of the land by wasteful slash-and-burn methods, by which trees are cut and set ablaze and their ashes used to fertilize the thin, rocky soil. A year or two of improper farming and grazing practices wear out the soil, and farmers are forced to abandon their fields in search of more land to clear. The abandoned fields are then subjected to severe soil erosion because plants no longer exist to protect against the effects of rain and wind. Once the soil disappears, rain forests that have been in existence for as long as 30 million years cannot return.

Developing countries clear the tropical rain forests on an unprecedented scale for cattle grazing, which tends to destroy the land. The beef is mostly exported to more affluent nations at relatively low prices. After a couple years of extensive agriculture, the soil is robbed of its nutrients. Because most of the world's farmers cannot afford expensive fertilizers, they are forced to abandon the infertile land. The denuded land is then subjected to severe soil erosion, often leaving behind bare bedrock.

Most African farmers are too poor to buy chemical fertilizers. The animal dung once used to enrich the soil is instead burned for fuel because forests have been cleared and supplies of firewood have dwindled. Moreover, deforestation causes the soil to lose much of its capacity to retain moisture, thereby reducing productivity and resistance to drought. Therefore, famine in Africa is becoming more of a human-caused disaster.

The destruction of the rain forests brings changing weather patterns within the forests themselves, converting woodlands into deserts. The continuing destruction of the world's forests also dramatically reduces their ability to absorb excess carbon dioxide generated by industrial activities. As the climate warms, some areas in the Northern Hemisphere would dry out, making ideal conditions for massive forest fires. If the warming continues, major forest fires might become more frequent, with substantial losses of forests and wildlife habitats.

The denuding of the world's forests increases the Earth's albedo with a consequential loss of precipitation. Soot from massive forest fires, especially in the Amazon Basin (Fig. 121), absorbs sunlight. This process heats the atmosphere and produces a temperature imbalance, causing temperatures to rise with altitude, just the opposite of what they should do. Therefore, large quantities of atmospheric soot generated by tropical forest fires could result in abnormal weather throughout the world.

Figure 121 *A smoke cloud from forest fires that obscured almost a third of South America in the fall of 1989.*

(Courtesy NASA)

The changing weather patterns put an additional strain on the forests and subjects them to infestation and disease, causing more trees to die. Furthermore, under normal conditions, high evaporation rates and transpiration rates within the forests help create more clouds, which contribute to the prodigious amounts of rainfall these areas receive. However, when people remove the forests, the cycle is broken. The resulting torrential rains cause severe flash floods that can be disastrous for those living along streams.

Severe erosion caused by large-scale deforestation can overload rivers with sediment, causing considerable problems downstream. Africa has the worst soil erosion in the world. As a result, its rivers are the most heavily polluted with sediment, whereas other rivers have completely dried out. Monsoon floodwaters cascading down the denuded foothills of the Himalayas of northern India and carried to the Bay of Bengal by the Ganges and Brahmaputra Rivers has devastated Bangladesh, where several thousand people have lost their lives to the floods. The Amazon River in South America is forced to carry much more water during the flood season due to deforestation at its headwaters. Deforestation and soil erosion are also causing many of the world's rivers to carry a higher sediment load.

DESERT EROSION

Heavy downpours along small drainage areas cause erosion of desert mountain ranges. Sediment fans consisting of sands and gravels develop at the mountain front. When the formally steep mountain front retreats, it leaves a smooth surface in the bedrock called a pediment. This generally has a concave-upward slope of up to 7 degrees, depending on the sediment size and the amount of runoff. Streams issuing from the mountains change course back and forth across the pediment in a manner similar to the formation of alluvial fans. Eventually, the mountains erode down to the level of the plain, leaving the pediments speckled with remnants of the range.

The development of drainage patterns in desert lands is well demonstrated in the Basin and Range Province of the American Southwest. The region contains several mountain ranges formed by relatively recent faulting. The basins between ranges are low-lying areas that often contained lakes during wetter climates. Lake-deposited sediments are common, and dry lake beds called playas cover the surface. The bodies of water are called alkali lakes because of their high concentrations of salt and other soluble minerals. When the lakes evaporate, they become alkali flats such as those in Utah's Great Salt Lake Desert (Fig. 122).

Only in desert and semiarid regions is wind an active agent of erosion, transportation, and deposition. The deserts host some of the strongest winds due to rapid heating and cooling of the land surface. The winds generate sandstorms and dust storms, which work together to cause wind erosion. Wind erosion develops mainly by deflation, which is the removal of large amounts of sediment by windstorms, forming a deflation basin. Deflation usually occurs in arid regions and unvegetated areas such as deserts and dry lake beds. In some areas, deflation produces hollows called blowouts, which are recognized by their typically concave shape. As smaller soil particles blow away during dust storms, the ground coarsens over time. The remaining sand tends to roll, creep, or bounce with the wind until it meets an obstacle, whereupon it settles and builds into a dune.

Figure 122 The south end of the Great Salt Lake Desert from southwest of Simpson Range, Tooele County, Utah in 1903.

(Photo by C. D. Walcott, courtesy USGS)

Often, after the finer material is evacuated, a layer of pebbles remains behind to protect against further deflation. Over a period of thousands of years, deserts develop a protective shield of pebbles coated with desert varnish, composed of magnesium and iron oxides exuded from the rock. The pebbles become fairly well embedded and vary from the size of a pea to the size of a walnut, making them too heavy for the strongest desert winds to pick up. This process helps to hold down the sand grains to create a stable terrain. Any disturbance on the surface can spawn a new generation of roving sand dunes and a higher incidence of dust storms that sweep the sediments from place to place.

Abrasion is similar to sandblasting by wind-driven sand grains that can cause erosion near the base of a cliff. When acting on boulders or pebbles, abrasion pits, etches, grooves, and scours exposed rock surfaces (Fig. 123). Abrasion also produces some unusually shaped rocks called ventifacts, which often have several flat, polished surfaces depending on the wind direction or the movement of the rock. Sandblasting erodes the surface of boulders, while desert varnish colors them dark brown or black. Maximum erosion effects occur during strong sandstorms, with sediment grains generally rising less than 2 feet above the ground. These abrasive effects occur most commonly on fence posts and power poles.

Sands act partly as solids and partly as liquids. Sand grains march across the desert floor under the influence of strong winds by a process called saltation (Fig. 124). The grains of sand become airborne for an instant, rising a foot or more above the ground. When landing, they dislodge additional sand

Figure 123 *Granite hollowed out by windblown sand, Atacama, Chile.*

(Photo by K. Segerstrom, courtesy USGS)

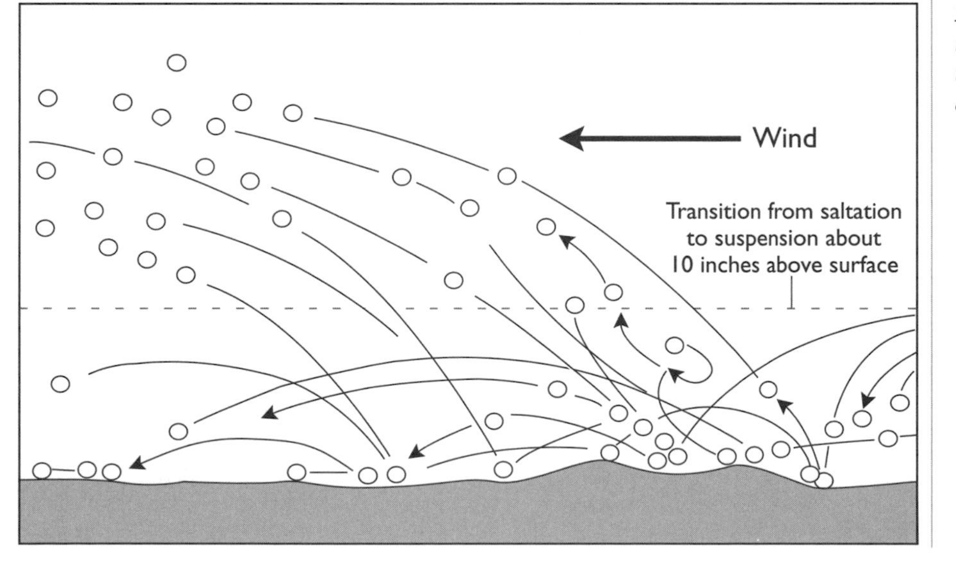

Figure 124 *The transition from saltation to suspension occurs about a foot above the surface.*

grains, which repeat the process. The rest of the moving sand travels forward along the surface by rolling and sliding. The sediment grains of desert deposits often have an opaque and frosted appearance due to abrasion from the constant motion of the sand.

HABOOBS

A severe dust storm that forms over a desert during convective instability, such as a thunderstorm (Fig. 125), is called a haboob, an Arabic word meaning "violent wind." Dust storms arise frequently in the Sudan of northern Africa where, in the vicinity of Khartoum, they are experienced about two dozen times a year. They are associated with the rainy season and remove a remarkable amount of sediment. A typical dust storm 300 to 400 miles in diameter can airlift more than 100 million tons of sediment. During the height of the season, between May and October, from 12 to 15 feet of sand can pile up against any obstruction exposed to the full fury of the storms.

Many severe dust storms occur in the American Southwest, especially near Phoenix, Arizona (Fig. 126), which averages about a dozen per year.

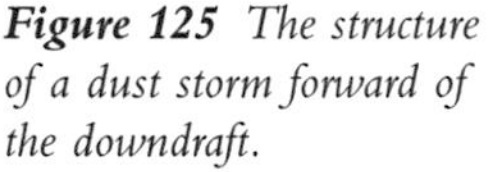

Figure 125 *The structure of a dust storm forward of the downdraft.*

Figure 126 *A massive dust storm rises over Phoenix, Arizona, on Labor Day, 1972.*
(Courtesy NOAA)

Often, people and animals die during severe dust storms. For instance, in 1895, a major dust storm in eastern Colorado is reported to have killed 20 percent of the cattle in the region. As in the Sudan, American dust storms occur most frequently during the rainy season, normally July and August. Surges of moist tropical air from the Pacific rush up from the Gulf of California into Arizona and generate long, arching squall lines, with dust storms fanning out in front.

Individual outflows often merge to form a solid wall of sand and dust that stretches for hundreds of miles. The sediment rises 8,000 to 14,000 feet above ground level and travels at an average speed of 30 miles per hour, with gusts of 60 miles per hour or more. Dust storms can also give rise to small, short-lived, and intense whirlwinds within the storms themselves or a short distance out in front that can damage buildings and other structures in their paths.

Average visibility falls to about a quarter of a mile but can drop to zero in very intense storms. After the storm ceases, an hour or so elapses before the skies began to clear and visibility returns to normal. However, if the parent thunderstorm arrives behind the dust storm, its precipitation clears the air much sooner. As often happens, however, the trailing thunderstorm fails to arrive or the precipitation evaporates before reaching the ground, a phenomenon known as virga. This causes the sediment to remain suspended for several hours or even days.

Vast dust storms occur when an enormous airstream moves across deserts, particularly those in Africa. Giant dust bands 1,500 miles long and 400 miles wide often traverse the region driven by strong cold fronts. Some large African storm systems have even carried dust across the Atlantic Ocean to South America, where about 13 million tons land in the Amazon basin annually. The dust over African deserts rises to high altitudes, where westward-flowing air currents transport it across the Atlantic. Fast-moving storm systems in the Amazon rain forest pull in the dust, which contains nutrients that enrich the soil.

So much African dust blows across the Atlantic during summer storms that millions of tons blanket Florida's skies and coat cars and other objects with a fine, red powder. Other areas along the East Coast of the United States actually violate clean air standards because of the additional load of African dust. Dust from the Sahara Desert blows across the rest of the United States, possibly reaching as far as the Grand Canyon, contributing the notorious haze that obscures the canyon's beauty. The dust is chemically different from local soils and has a distinctive red-brown color. When added to other air pollutants, the Sahara dust causes a persistent haze, especially in summer.

The dust has an unexpected benefit, however. The periodic influxes of calcium-rich sediment help regions plagued with acid rain produced when fossil fuels are burned by diluting the acidic content of rainwater. The dust supplies the ocean with much of its iron, an important nutrient needed to keep the marine ecosystem healthy. Coral off the Florida Keys trap the dust inside growth bands, which can be used to trace dust from sources such as storms of sand blowing off the Sahara Desert toward the United States.

Massive dust storms also arise in Arabia, central Asia, central China, and the deserts of Australia and South America, where the most obvious threat is soil erosion. The strong winds create gigantic dust storms that cause severe erosional problems. The tendency of the wind to erode the soil is often aggravated by improper agricultural practices. Wind erosion takes out of production an estimated 1.2 million acres of farmland in Russia annually, increasing the difficulty for the nation to feed itself. In the United States, wind erodes about 20 million tons of soil per year. The primary method of controlling wind erosion is maintaining a surface cover of vegetation. However, if rainfall is insufficient, these measures often fail, and the soil simply blows away.

In dry regions where dust storms were once prevalent, the wind transported large quantities of loose sediment. These sediments accumulated into thick deposits of loess (Fig. 127), covering thousands of square miles. Loess is a fine-grained, loosely consolidated, sheetlike deposit that often shows

thin, uniform bedding on outcrop. Secondary loess deposits were transported and reworked over a short distance by water or intensely weathered in place.

Figure 127 *Exposure of loess in vertical cliffs, Warren County, Mississippi, in 1915.* (Photo by E. W. Shaw, courtesy USGS)

The sediment comprises angular particles of equal-sized grains. The particles are composed of quartz, feldspar, hornblende, mica, and bits of clay. It is usually a buff to yellowish brown loamy deposit that is commonly unstratified due to a rather uniform grain size, generally in the silt size range. Loess often contains the remains of grass roots. Like mud bricks, deposits can stand in nearly vertical walls despite their weak cohesion. Loess can also cause problems in construction unless properly compacted because on wetting, it tends to settle.

Loess is common in North America, Europe, and Asia, with China containing the world's largest deposits. They originated from the Gobi Desert and attain hundreds of feet in thickness. Most loess deposits in the central United States are located adjacent to the Mississippi River valley, where nearly a quarter million square miles are covered by sediment from the glaciated northlands. Deposits also cover portions of the Pacific Northwest and Idaho. Loess makes a yellowish fertile soil responsible for much of the abundant agricultural production of the American Midwest.

SAND DUNES

About 10 percent of the world's deserts are composed of sand dunes. They are driven across the desert by the wind and engulf everything in their paths, including man-made structures. Sand dunes also pose a major problem in the construction and maintenance of highways and railroads that cross sandy areas of deserts. Sand dune migration near desert oases poses another serious problem, especially when encroaching on villages. Methods to mitigate damages to structures from sand dunes include building windbreaks and funneling sand out of the way. Without such measures, disruption of roads, airports, agricultural settlements, and towns could become a major problem in desert regions.

The direction, strength, and variability of the wind, the moisture content of the soil, the vegetation cover, the underlying topography, and the amount of movable soil exposed to the wind determine the size and form of sand dunes. They generally have three basic shapes, determined by the topography of the land and patterns of wind flow. Linear dunes (Fig. 128) are aligned in roughly the direction of strong, steady, prevailing winds. Their length is substantially greater than their width, and they lie parallel to each other, sometimes with a wavy pattern. When the wind blows over the dune peaks, part of the air flow shears off and turns sideways. The air current scoops up sand and deposits it along the length of the dune, which maintains its height and simultaneously lengthens it. The surface area covered by dunes is about equal to the area between dunes. Both sides of the dune are likely steep enough to cause landslides.

Crescent dunes, also called barchans, are symmetrically shaped with horns pointing downwind. They travel across the desert at speeds of up to 50 feet a year. Parabolic dunes form in areas where sparse vegetation anchors the side arms, while the center is blown outward, causing sand in the middle to move forward. Star dunes (Fig. 129) form by shifting winds that pile up sand into central points that can rise 1,500 feet and more, with several arms radiating outward, looking much like giant pinwheels. Sand also accumulates in flat sheets or forms stringers downwind that do not exhibit any appreciable relief in sand seas.

A curious feature exhibited by sand dunes is an unexplained phenomenon known as booming sands. At least 30 booming dunes have been found in deserts and on beaches in Africa, Asia, North America, and elsewhere. The sound occurs almost exclusively in large, isolated dunes deep in the desert or on back beaches well inland from the coast. The noises can be triggered by simply walking along the dune ridges. When sand slides down the lee side of a dune, it sometimes emits a loud rumble. The sounds emitting from the dunes

Figure 128 *A Skylab photograph of linear dunes in the northwest Sahara Desert of North Africa.*

(Photo by E. D. McKee, courtesy USGS)

have been likened to bells, trumpets, pipe organs, foghorns, cannon fire, thunder, buzzing telephone wires, or low-flying aircraft.

The grains in sound-producing sand are usually spherical, well rounded, and well sorted or of equal size. The dune is usually far from its original sand source so that winds carry grains along for great distances, thereby depositing

Figure 129 *Compound star dunes in Gran Desierto, Sonora, Mexico.* (Photo by E. D. McKee, courtesy USGS)

similarly sized, well-rounded grains at or near the top of the dune. In addition, the wind must be strong enough to push sand over the top, causing it to avalanche downslope. However, sands packed too tightly or too loosely do not shear properly to produce the right sounds. The sound appears to originate from a harmonic event occurring at the same frequency. Normal landsliding, however, involves a mass of randomly moving sand grains that collide with a frequency much too high to produce such an unusual noise.

DUST BOWLS

Evidence suggests the American Great Plains experienced tremendous dust storms long before the introduction of agriculture in the region. The area has suffered repeated droughts for thousands of years. Dry spells not only persisted for centuries at a time but occurred much more frequently than they do today. The worst drought years were A.D. 200–270, 700–850, and 1000–1200. The Anasazi who occupied cliff dwellings at Mesa Verde (Fig. 130) and elsewhere in the American Southwest, mysteriously disappeared during the mid-13th century, possibly due to a prolonged drought, because they relied heavily on farming. Modern farming and ranching methods have seriously aggravated the problem. Over the last 150 years, the average soil depth in the most productive areas of the United States has been cut in half by intensive agriculture.

Droughts are periods of abnormally dry weather resulting from shifting precipitation patterns around the world. They can be sufficiently prolonged for the lack of water to undermine agriculture seriously. In 1983, the United States experienced one of the worst droughts since the Dust

Figure 130 *Ruins of the Anasazi Mesa Verde National Park, Colorado.* (Courtesy National Park Service)

Bowl of the 1930s, with crop loses in the billions of dollars. The 1983 drought in Australia was the most severe in more than 100 years. The nearly year-long drought cut grain production by about half that of the previous year. Thousands of sheep and cattle dying of starvation and thirst had to be destroyed and buried in mass graves. An equally intense drought caused food shortages in southern Africa and affected West Africa and the Sahel region bordering the Sahara Desert. The sub-Saharan drought of the last quarter century was the worst in 150 years. The 1983 and 1984 droughts, which left upward of a million people dead or dying of famine, were the worst of the 20th century.

The droughts might have been triggered by an unusually warm tropical Pacific during an El Niño event and its accompanying atmospheric changes. The Pacific warming often reduces rainfall in Australia, Indonesia, parts of Brazil, and eastern and southern Africa and increases rainfall in the normally dry west coast of South America and the Pacific Coast of North America. El Niños are anomalous warming conditions in the eastern equatorial Pacific Ocean due to the failure of the westerlies, the eastward-blowing trade winds.

El Niños have historically recurred once every three to seven years and last for up to two years. However, in the last 20 years, they have become stronger and longer. The strongest El Niño on record registered Pacific water temperatures as high as 5 degrees Celsius above normal from 1997 to 1998. The higher frequency of El Niño events appear to be a symptom of greenhouse gas pollution and global warming. By contrast, a colder-than-normal tropical Pacific known as La Niña tends to dry out the southwestern and south-central United States, resulting in high summer temperatures and drought conditions.

Increasing surface temperatures could have an adverse effect on global precipitation, with some areas receiving less rainfall and others more. This would change global occurrences of droughts and floods. Areas lying 30 degrees on either side of the equator can expect dramatic shifts in precipitation patterns as the world continues to warm. The seasonal winds of the monsoons, which bring much-needed rainfall to half the people of the world, affect the continents of Asia, Africa, and Australia. However, perturbations in the climate induced by climate change could lead to years of drought or flood, placing multitudes of people into great peril.

Changes in precipitation patterns (Fig. 131) could have a profound effect on the distribution of water resources. Higher temperatures would augment evaporation, causing the flow of some rivers to decline by as much as 50 percent or more, while other rivers might dry out entirely. During the 1988 drought in the United States, the Mississippi River fell to a record low, making navigation impossible over long stretches. Ancient sunken derelicts were

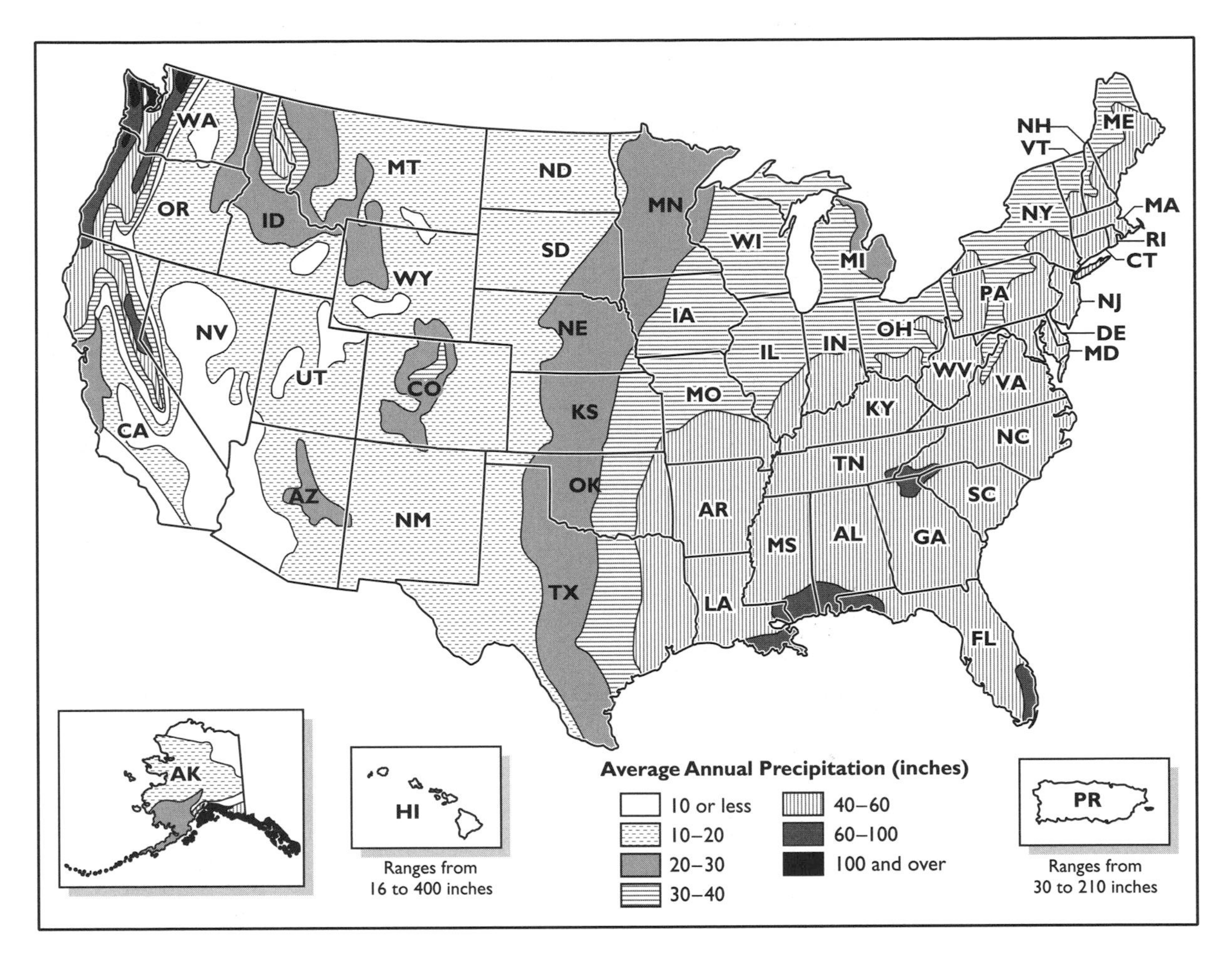

Figure 131 *Annual rainfall in inches in the United States.*

exposed for the first time in the century, underscoring the severity of the drought.

Global climate change can potentially increase the frequency and severity of droughts, with continental interiors that experience occasional droughts becoming permanently dry wastelands. Soils in almost all of Europe, Asia, and North America could become drier, requiring up to 50 percent additional irrigation. Rises in temperatures, increases in evaporation rates, and changes in rainfall patterns would severely limit the export of excess food to developing countries during times of famine.

Subtropical regions might experience a marked decrease in precipitation, encouraging the spread of deserts. Increasing the area of desert and semidesert regions would significantly affect agriculture, which would be forced to move to higher latitudes. Canada and Russia would then turn into breadbaskets, while the United States might have to import grain. Unfortunately, the

Figure 132 *Soil drifting from an unprotected corner of a wheat field in Chase County, Nebraska.*

(Photo by H. E. Alexander, courtesy USDA Soil Conservation Service)

soils in the northern regions are thin due to glacial erosion during the last Ice Age and would soon wear out from extensive agriculture. Furthermore, changing weather patterns due to instabilities in the atmosphere could make deserts out of once productive farmlands (Fig. 132). Dry winds of tornadic force could create gigantic dust storms and serious erosional problems as much of the land becomes a hot, dry desert.

After seeing the effects of sand in motion, the next chapter reveals the geologic hazards of glacial ice on the run during the Ice Age and at present, including the affects of glacial surge and rising sea levels due to the melting of the world's ice caps.

8

GLACIERS

ICE ON THE RUN

This chapter examines the geologic hazards posed by glaciers in motion and the melting of the great ice caps. Glacial ice covers approximately one-tenth of the Earth's surface, and glaciers contain about 70 percent of all the Earth's freshwater. Most of the world's ice lies atop Antarctica, and smaller ice caps cover Greenland and Iceland. Alpine glaciers are found on every continent and hold as much freshwater as all the world's rivers and lakes.

If the ice caps melt during a sustained warmer climate, they could substantially raise sea levels and drown coastal regions. Beaches and barrier islands would disappear as shorelines continue to move inland (Fig. 133). Half the people of the world live in coastal areas. Rich river deltas that feed a large portion of the world's population would be inundated by the rising waters. Coastal cities would have to move farther inland or build costly seawalls to keep out the raging sea.

Figure 133 *Development along the shorefront of Ocean City, Maryland. The distances between these buildings and the shoreline leaves little room for natural processes during storms.*

(Photo by R. Dolan, courtesy USGS)

GLACIATION

The northern lands owe their unusual landscapes to massive ice sheets that swept down from the polar regions during the Ice Age. In places, glaciers scraped the crust completely clean of sediments, exposing the raw basement rock below. In other areas, glaciers deposited huge heaps of sediment (Fig. 134) when they retreated to the poles. Perhaps within another couple thousand years or so, the ice sheets will again be on the rampage, wiping out everything in their paths. Rubble from northern cities would be bulldozed hundreds of miles to the south.

During the Ice Age, huge masses of ice spread outward from the polar regions. Glaciers up to 2 miles or more thick enveloped Canada, Greenland, and northern Eurasia. The glaciers covered some 11 million square miles of land that is presently ice free. The glaciation began with a rapid buildup of glacial ice some 115,000 years ago, intensified about 75,000 years ago, and peaked about 18,000 years ago (Table 8).

North America was engulfed by two main glacial centers. The largest ice sheet, called the Laurentide, blanketed an area of about 5 million square miles.

Figure 134 *Grass-covered glaciated topography, Glacier County, Montana.*

(Photo by H. E. Malde, courtesy USGS)

TABLE 8 THE MAJOR ICE AGES

Time (in years)	Event
12,000–present	Present interglacial
16,000–12,000	Melting of ice sheets
20,000–18,000	Last glacial maximum
100,000	Most recent glacial episode
1 million	First major interglacial
3 million	First glacial episode in Northern Hemisphere
4 million	Ice covers Greenland and the Arctic Ocean
15 million	Second major glacial episode in Antarctica
30 million	First major glacial episode in Antarctica
65 million	Climate deteriorates, poles become much colder
250–65 million	Interval of warm and relatively uniform climate
250 million	The great Permian ice age
700 million	The great Precambriann ice age
2.4 billion	First major ice age

It extended from Hudson Bay and reached northward into the Arctic Ocean and southward into eastern Canada, New England, and the upper midwestern United States. It was apparently unstable and partially collapsed several times during the last Ice Age, spawning armadas of icebergs into the Atlantic Ocean.

A smaller ice sheet, called the Cordilleran, originated in the Canadian Rockies and enveloped western Canada and the northern and southern sections of Alaska, leaving an ice-free corridor down the center of the state. Scattered glaciers also covered the mountainous regions of the northwestern United States. Ice buried the mountains of Wyoming, Colorado, and California, and rivers of ice linked the North American cordillera with mountains in Mexico.

Two major ice sheets overran parts of northern Europe as well. The largest, called the Fennoscandian, fanned out from northern Scandinavia and covered most of Great Britain as far south as London and large parts of northern Germany, Poland, and European Russia. A smaller ice sheet, known as the Alpine, was centered in the Swiss Alps and enveloped parts of Austria, Italy, France, and southern Germany. In Asia, glaciers occupied the Himalayas and blanketed parts of Siberia.

In the Southern Hemisphere, only Antarctica held a major ice sheet, which expanded to about 10 percent larger than its present size and extended as far as the tip of South America. Sea ice surrounding Antarctica nearly doubled its modern wintertime area. Smaller glaciers capped the mountains of Australia, New Zealand, and the Andes of South America, which contained the largest of the southern alpine ice sheets. Throughout the rest of the world, mountain glaciers topped peaks that are currently ice free.

Approximately 5 percent of the planet's water was locked up in glacial ice. The continental ice sheets contained some 10 million cubic miles of water and covered about one-third the land surface with glacial ice three times its current size. The accumulated ice dropped sea levels about 400 feet, and shorelines advanced seaward up to 100 miles or more. The drop in sea level exposed land bridges and linked continents, spurring a vigorous migration of species, including humans, to various parts of the world.

Precipitation rates fell because the lower temperatures reduced the evaporation rate of seawater and decreased the average amount of snowfall. Since little melting took place during the cooler summers, only minor amounts of snowfall were required to sustain the ice sheets. The lower precipitation levels also increased the spread of deserts across many parts of the world. Fierce desert winds produced tremendous dust storms. The dense dust suspended in the atmosphere blocked sunlight, keeping temperatures well below present-day averages.

Perhaps one of the most dramatic climate changes in geologic history took place when the Ice Age ended about 10,000 years ago during the pre-

Figure 135 *An erratic boulder southeast of Sentinel Dome, Yosemite National Park, in 1913.*
(Photo by F. C. Calkins, courtesy USGS)

sent interglacial known as the Holocene epoch. After some 100,000 years of gradual accumulation of snow and ice up to 2 miles and more thick, the glaciers melted away in just a few thousand years, withdrawing several hundred feet annually. The retreating glaciers left an assortment of glacial deposits, including sinuous eskers, elongated drumlins, and immense boulder fields (Fig. 135).

About a third of the ice melted between 16,000 and 12,000 years ago when average global temperatures increased about 5 degrees Celsius to nearly present levels. The reestablishment of the deep-ocean circulation system that pumps warm water to the higher latitudes, which was shut off or severely weakened during the Ice Age, might have thawed out the planet from its deep freeze. Then between 11,000 and 10,000 years ago, during a period known as the Younger Dryas, named for an Arctic wildflower, the glaciers paused in midstride as temperatures suddenly fell again to ice age levels. Afterward, a second episode of melting ensued, which led to the present volume of ice by about 6,000 years ago.

The most recent Ice Age is the best studied of all previous glaciations because their evidence was erased, as ice sheets eradicated much of the landscape. In many areas, the ice stripped off entire layers of sediment, leaving bare bedrock behind (Fig. 136). In other areas, older deposits were buried under thick glacial deposits.

Most of the evidence for extensive glaciation is found in moraines and tillites, which are glacially deposited rocks. Moraines are accumulations of rock material carried by a glacier and deposited in a regular, usually linear pattern that makes a recognizable landform. Many parts of the world overrun by thick glaciers during the last Ice Age were eroded down to the granite bedrock, and the debris was left in great heaps. The glacially derived sediments covered much of the landscape, burying older rocks under thick layers of till. Tillites, a mixture of boulders and pebbles in a clay matrix consolidated into solid rock, were deposited by glacial ice and are known to exist on every continent.

THE POLAR ICE CAPS

Within the last few million years, permanent ice covered both poles. This was a unique event in geologic history because just having a single polar ice cap was rare. The formation of glacial ice resulted from the configuration of the Earth's crust. In our planet's early history, continents converged on regions

Figure 136 *A remnant of a preglacial erosion surface that forms the top of Dora Mountain at an altitude of 12,300 feet. Dora Lake in the foreground is dammed at left by low early glacial moraine from an icecap, Summit County, Colorado.*

(Photo by O. Tweto, courtesy USGS)

around the equator, allowing warm ocean currents access to the polar regions, which kept them ice free year-round. During the last 100 million years, however, the continents shifted positions, so that a large continent covered the South Pole and a nearly landlocked sea surrounded the North Pole.

Most of the continental landmasses moved north of the equator, leaving little land in the Southern Hemisphere, where ocean spans nine-tenths of the surface. The drifting of the continents radically changed patterns of ocean currents, whose access to the poles was restricted. Without warm ocean currents flowing from the Tropics to keep the polar regions free of ice, glaciers will remain until such a time when the continents once again drift back toward the equator, perhaps in another 100 million years.

Greenland, the world's largest island, drifted apart from Eurasia and North America about 60 million years ago. Occasionally, Alaska and Siberia connected and closed off the Arctic Basin, isolating it from warm-water currents originating in the Tropics resulting in the accumulation of pack ice. About 40 million years ago, Antarctica separated from Australia and drifted over the South Pole, where it acquired a continent-sized ice sheet (Fig. 137). The existence of ice at both poles established a unique equator-to-pole oceanic and atmospheric circulation system.

The Tethys Sea was a large, shallow, equatorial body of water separating Africa and Eurasia during the Mesozoic and early Cenozoic eras. Warm water, top-heavy with salt from high evaporation rates and little rainfall, sank

Figure 137 *Daniell Peninsula, Antarctica.*

(Photo by W. B. Hamilton, courtesy USGS)

to the bottom of the sea. Meanwhile, Antarctica, whose climate was much warmer than today, generated cool water that filled the upper layers of the ocean, causing the entire ocean circulation system to run backward. About 28 million years ago, Africa collided with Eurasia, which turned off warm-water currents flowing to the poles. This resulted in the formation of a major ice sheet in Antarctica. Cold air and ice flowing into the surrounding sea cooled the surface waters. The cold, heavy water sank to the bottom and flowed toward the equator, generating the ocean circulation system that exists today.

Some 3 percent of the planet's water is locked up in the polar ice caps, which cover on average about 7 percent of the Earth's surface. The Arctic is a sea of pack ice, whose boundary is the 10-degree Celsius July isotherm, the extent of polar drift ice during summer. A permanent ice cap did not develop over the North Pole until about 8 million years ago, when Greenland acquired its first major ice sheet. The sea ice covers an average area of about 4 million square miles, with an average thickness of 15 to 20 feet. The Arctic pack ice often aggregates into unusual shapes, including rounded forms 2 feet or more in diameter called pancake ice, making the sea resemble an unending pond of icy water lilies.

By far, the greatest amount of ice lies atop Antarctica, which covers about 5.5 million square miles, an area larger than the United States, Mexico, and Central America combined. Entire mountain ranges are covered by a sheet of ice 3 miles thick in places, with an average thickness of 1.3 miles and a mean elevation of about 7,500 feet above sea level. The total volume of Antarctic ice is approximately 7 million cubic miles, enough to make an enormous ice cube nearly 200 miles on a side.

About 37 million years ago, global temperatures plummeted due to continental movements. At this time, Antarctica accumulated a thick layer of ice that dwarfed even the present ice sheet. Sometime during the following 15 million years, most of the ice sheet melted, probably due to a warmer global climate. Around 13 million years ago, a new ice sheet formed as the climate grew colder and ocean bottom temperatures approached the freezing point. All the continent's land features, including canyons, valleys, plains, plateaus, and mountains, were buried under ice.

The snows in Antarctica accumulate into thick ice sheets because virtually no melting takes place from year to year. The summer mean monthly temperature is –35 degrees Celsius, and the winter mean monthly temperature is –60 degrees in places dropping to –90 degrees. Barren mountain peaks soar 17,000 feet above the ice sheet, and 200-mile-per-hour winds shriek off the ice-laden mountains and high ice plateaus. The Transantarctic Range divides the continent into a large eastern ice mass and a smaller western lobe about the size of Greenland. The West Antarctic ice sheet apparently did not form

Figure 138 *Sea ice along with a large iceberg in the Antarctic Sea.*

(Courtesy U.S. Maritime Administration)

until about 9 million years ago and lies mostly on the continental shelf anchored by scattered islands.

During the winter months, from June to September, nearly 8 million square miles of ocean that surrounds Antarctica is covered by sea ice, with an average thickness of usually no more than 3 feet (Fig. 138). As temperatures plummet, sea ice begins growing at an average rate of some 20 square miles per minute. Because of this great expanse of ice, Antarctica plays a more significant role in atmospheric and oceanic circulation than does the Arctic. The sea ice is punctured in various places by coastal and ocean polynyas, which are large, open-water areas kept from freezing by upwelling warm-water currents. The coastal polynyas are essentially sea ice factories because they expose portions of open ocean that later freeze, thereby continuing the ice-manufacturing process.

The waters surrounding Antarctica are the coldest in the world. A thermal barrier produced by the circum-Antarctic current impedes the inflow of warm ocean currents. The sea ice covering the ocean around Antarctica remains for at least 10 months of the year, during which time the continent is in near total darkness for 4 months. The water temperature throughout the year varies from 2 to 4 degrees below freezing. However, due to its high salt content, seawater does not freeze. Neither do Antarctic fish, which contain an antifreeze-like substance in their bodies that keeps them alive during the cold winter months.

CONTINENTAL GLACIERS

Continental glaciers are the world's largest ice sheets. During the last Ice Age from about 115,000 to 12,000 years ago, ice sheets covered one-third of the land surface. Today, only Antarctica and Greenland contain substantial ice masses, with about 30 percent of the total ice volume that existed during the last Ice Age. A continental glacier moves in all directions outward from its point of origin. It completely engulfs the land except for isolated high mountain peaks projecting above the surface of the ice. The term ice cap also describes a small glacier that spreads out radially from a central point as on Iceland.

At the margins of the ice sheets are rugged periglacial deposits sculpted from solid bedrock. These features developed along the tip of the ice and were directly controlled by the glacier. Cold winds blowing off the ice sheets affected the climate of the glacial margins and helped create periglacial conditions. The zone was dominated by such processes as frost heaving, frost splitting, and sorting, which created immense boulder fields out of what was once solid bedrock.

The largest ice sheet lies atop the continent of Antarctica. It has geographic features similar to those of other continents, except its mountain ranges, high plateaus, lowland plains, and canyons are buried under a thick sheet of ice (Fig. 139). The ice on the mainland is unimaginably heavy and

Figure 139 *A volcano rising above the glacial ice in Antarctica.*

(Photo by W. B. Hamilton, courtesy USGS)

depresses the continental bedrock nearly 2,000 feet. The ice cap is so thick the region is practically devoid of earthquakes because the great weight of the ice prevents slippage along faults. Interestingly, at the end of the Ice Age, when the great northern glaciers melted and the crust rebounded as the weight of the ice was removed, huge earthquakes struck the once icebound regions.

Antarctica is one of the driest regions in the world. It has an average annual snowfall of less than 2 feet, which translates into roughly 3 inches of rain. However, because little melting occurs in the continental interior, any new snowpack adds to the growth of the ice sheets. Only about 2.5 percent of this huge landmass is free of ice for a few months of the year. Dry valleys (Fig. 140) gouged out by local ice sheets running between McMurdo Sound and the Transantarctic Mountains are the largest ice-free areas on the continent. They receive less than 4 inches of snowfall yearly, most of which blows away by strong winds. The landscapes of the dry valleys are very old. Some surfaces are quite steep and appear to have remained virtually unchanged for 15 million years.

Nine-tenths of the Earth's ice lies atop Antarctica, whose glaciers contain 70 percent of all the world's freshwater. The bulk of the ice has not changed significantly over the past several million years, when the Earth took a plunge into a colder climate. Trapped under the thick polar ice cap is a huge expanse of water the size of Lake Ontario, covering more than 5,000 square miles to a depth of at least 1,600 feet. Water collected in bedrock pockets was prevented from freezing by geothermal heat from below and pressure from the ice above. The lake might even be inhabited by strange microbes, making this one of the harshest ecosystems on Earth. Sediments deposited onto the lake bottom by a glacier as it grinds over the bedrock would provide a significant source of nutrients.

Antarctica is surrounded by sea ice, which expands to about 7.7 million square miles in winter, more than twice the size of the United States. Antarctic sea ice differs from that of the Arctic, where most of the ocean is surrounded by land. The land dampens the seas by reducing the severity of storms and allows the ice to grow over twice as thick. Some Arctic ice survives the summer, so in 4 years it doubles in thickness. However, in the Antarctic, powerful storms at sea churn the water and break up the ice, preventing it from growing any thicker. If the apparent global warming continues, Antarctic ice sheets could become destabilized and break off into the ocean, creating additional sea ice. The increased area of sea ice could form a gigantic ice shelf, covering up to 10 million square miles.

About half of Antarctica is bordered by ice shelves (Fig. 141), which are thick sheets of floating ice that have slid off the continent. The continental ice sheet slowly flows down to the sea, where portions break off to form icebergs.

The floating ice shelves of the Ross and Weddell Seas dominate West Antarctica. The elevation in this region is generally low, and most of the ice rests on glacial till lying mostly below sea level. The till is a mixture of ground-up rock and water that acts as a lubricant to aid the ice sheet's slide into the ocean. The water pressure in the till below a moving ice stream is higher than that below a stationary one. The higher the pressure, the more easily the ice floats and therefore the more easily it moves.

Figure 140 *The upper part of Wright Dry Valley, Taylor Glacier region, Victoria Land, Antarctica.*

(Photo by W. B. Hamilton, courtesy USGS)

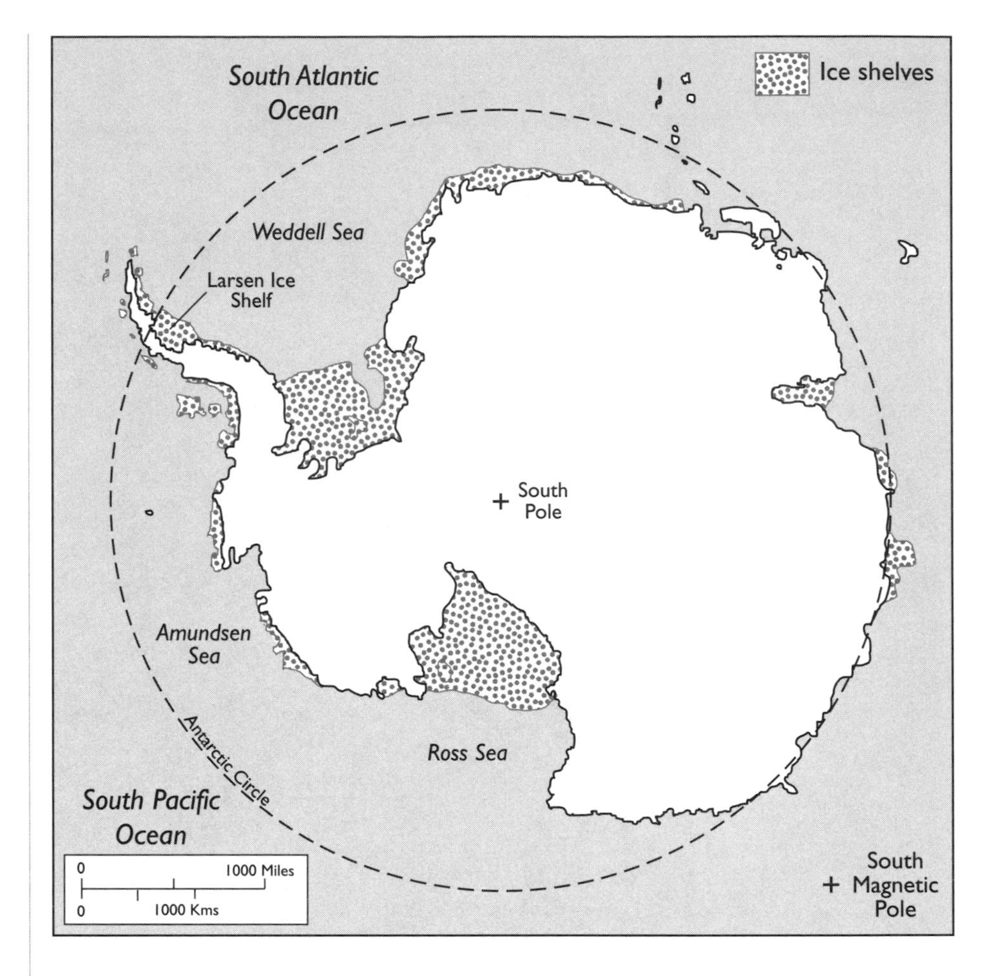

Figure 141 *The Antarctic ice shelves shown in stippled areas.*

The Filchner-Ronne Ice Shelf south of the Weddell Sea, the most massive floating block of ice on Earth, has two distinct layers. The top layer measures about 500 feet thick and is composed of ice mostly formed by falling snow. The bottom layer measures about 200 feet thick and consists of frozen seawater. The freshwater layer contains opaque and granular ice similar to the upper portion of a glacier. In contrast, the transparent marine shelf ice displays many inclusions of marine origin such as plankton and clay particles. Free-floating ice platelets recrystallize at the base of the marine layer, forming a slush that compacts into solid ice.

The largest ice sheet in the Northern Hemisphere overlies Greenland. About 8 million years ago, Greenland acquired a permanent ice cap that is, in some places, 2 miles or more thick. Snow precipitating from storm systems traversing across the surface of the glacier nourishes the Greenland ice sheet. The loss of ice at the boundary regions adjacent to the ocean balances the glacier's growth. Although Greenland is similar in size to West Antarctica, it has only small ice shelves in the far north. Large icebergs calving off glaciers entering the sea become a shipping hazard in the North Atlantic.

Iceland has the second largest ice cap in the Northern Hemisphere and is one of the coldest inhabited places on Earth. The island is a broad volcanic plateau of the Mid-Atlantic Ridge, which rose above the sea some 16 million years ago. The abnormally elevated topography extends some 900 miles along the ridge, 350 miles of which lies above sea level. South of Iceland, the broad plateau tapers off, forming a typical midocean ridge. A mantle plume rising from the bottom of the mantle and lying beneath the plateau apparently augments the normal volcanic flow of the Mid-Atlantic Ridge, making Iceland's existence possible.

Iceland is unique because it straddles a spreading-ridge system, where the two plates of the Atlantic Basin and adjacent continents pull apart. A steep-sided, V-shaped valley runs northward across the entire length of the island. It is one of the few expressions of a volcanic rift system on land, with many volcanoes flanking the rift. Volcanism on Iceland produces glacier-covered volcanic peaks up to a mile high and generates intense geothermal activity. Although Iceland is fortunate to possess an abundant supply of energy for electric power and heating, this energy source is not without its dangerous side effects. Frequent volcanic eruptions plague the island. The most destructive eruption in modern times buried much of the fishing village of Vestmannaeyjar on the island of Heimaey in 1973 (Fig. 142).

Figure 142 *Houses partially buried by tephra in the eastern part of Vestmannaeyjar, Iceland, during eruption on Heimaey on July 23, 1973.*

(Courtesy USGS)

Figure 143 *Damage to a bridge caused by a glacial outburst flood on Sheep Creek, east of Valdez, Bremner District, Copper River region, Alaska.*

(Photo by A. Post, courtesy USGS)

Perhaps the strangest volcanic eruption occurred under a glacier in 1918. It unleashed a massive flood of meltwater called a glacial burst or jokulhlaup, known to Icelanders since the 12th century. A glacier burst is a sudden release of meltwater from a glacier or subglacial lake that can be quite destructive (Fig. 143). Several underglacier eruptions have occurred in Iceland during the past century. Water gushing from such a glacier carves out an enormous ice cave. Geothermal heat beneath the ice creates a large reservoir of meltwater up to 1,000 feet deep, while a ridge of rock acts as a dam to hold back the water. When the dam suddenly breaks, the flow of water forms a channel under the ice that can extend up to 30 miles long or more.

An underglacier eruption in the sparsely populated southeastern part of Iceland on September 30, 1996, melted through the 1,700-foot-thick ice cap and sent massive floodwaters and icebergs dashing 20 miles to the sea a month later. In a matter of days, the below-glacier eruption released up to 20 times more water than the flow of the Amazon, the world's largest river. It destroyed telephone lines, bridges, and the only highway running along Iceland's southern coast.

The Antarctic ice sheets hide many volcanoes. Under-ice eruptions as much as a mile or more beneath the glacier can produce massive floods of meltwater. Antarctica is noted for its unusual volcanoes. The most active is Mount Erebus, a smoldering mountain that rises 12,500 feet above Ross Island. Several other volcanoes puncture the ice of West Antarctica. Many of Antarctica's dormant volcanoes are buried within the ice, and extensive volcanic deposits underlie the ice sheets. When active volcanoes erupt beneath the ice, they spout great floods of meltwater. The water mixes with underlying sediment, forming glacial till tens of feet thick. Basalt erupted beneath the glacial ice produces volcanic rocks called hyaloclastics. They are pillow lavas and pillow breccias, which are unique, quickly frozen forms of lava.

An active volcano erupting under the glacier melted a large area of ice, producing a round depression in the Ross Ice Shelf that measured roughly 4 miles wide and 160 feet deep. A 4-mile-wide, 2,100-foot-high volcano was found lying beneath more than a mile of ice. The volcano sits in the middle of a giant caldera 14 miles wide within a rift valley, where the Earth's crust is being stretched apart and hot rock from the mantle is rising to the surface.

Although this was the first active volcano found under the Antarctic ice, it is unlikely the only one. Other circular depressions in the ice suggest many more volcanoes are lurking beneath the glaciers. The volcanoes could produce enough heat to melt the base of the ice sheet, allowing ice streams tens of miles wide to flood into the sea, setting the stage for the collapse of the West Antarctic ice sheet. The surge of ice into the ocean would raise global sea levels, radically altering the landscape of the world.

ICE STREAMS

Large, flat areas beneath the Antarctic ice sheet are thought to be subglacial lakes, kept from freezing by the interior heat of the Earth. The temperature a mile below the surface of the ice can be warmer than the temperature of the ice on top. The high pressures that occur at such depths enable liquid water to exist several degrees colder than its normal freezing point. The pools of liquid water tend to lubricate the ice, enabling ice streams up to several miles broad to glide smoothly along the valley floors. The ice streams flow easily down the mountain valleys to the sea.

Every year, about a trillion tons of ice discharge into the seas surrounding Antarctica. As much as 90 percent of the discharge from the Antarctic ice sheet drains by a small number of fast-moving ice streams. The ice flowing into the sea often calves off to form icebergs (Fig. 144). Furthermore, the icebergs appear to be getting larger, possibly due to a warmer global climate. The number of extremely large icebergs has also increased dramatically.

One of the largest known icebergs separated from the Ross Ice Shelf in late 1987. It measured about 100 miles long, 25 miles wide, and 750 feet thick, about twice the size of Rhode Island. In August 1989, the iceberg collided with Antarctica and broke in two. Another extremely large iceberg measuring 48 miles by 23 miles broke off the floating Larson Ice Shelf in early March 1995 and headed into the Pacific Ocean. The northern portion of the Larson

Figure 144 *U.S. Navy ships move a huge iceberg from a channel of broken ice leading to McMurdo Station, Antarctica.*

(Photo by A. W. Thomas, courtesy U.S. Navy)

Ice Shelf, located on the east coast of the Antarctic Peninsula, has been rapidly disintegrating. This breakup accounts for such gargantuan icebergs that have plagued the Antarctic Sea.

Perhaps during the biggest ice-breaking event in a century, an iceberg about 180 miles long and 25 miles wide, or roughly the size of Connecticut, split off from the Ross Ice Shelf in early spring 2000. The breaking off of the iceberg is most likely part of the normal process of ice shelf growth and not necessarily a consequence of global warming. These giant icebergs could pose a serious threat if they drift into the Ross Sea and block shipping lanes to McMurdo Station 200 miles away.

Behind a wall of mountains that form the spine of Antarctica, called the Transantarctic Range, rivers of ice slowly flow outward and down to the sea on all sides. The ice escapes through mountain valleys to the ice-submerged archipelago of West Antarctica and to the great floating ice shelves of the Ross and Weddell Seas. West Antarctica is traversed by ice streams several miles broad, where rivers of solid ice flow down mountain valleys. The banks and midsections of the ice streams might contain muddy pools of melted water that allow the glaciers to flow along the valley floors and plunge into the sea. The amount of ice flowing to the coast is significantly greater than the quantity accumulating at the ice stream's source. This indicates a possible instability that warrants urgent study to understand the nature of ice streams and to predict their future behavior.

Deep crevasses mark the banks and interior portions of the ice streams (Fig. 145). A glacial crevasse is a crack or fissure in a glacier, resulting from stress due to movement in the ice flow. Crevasses are generally several tens of feet wide, a hundred or more feet deep, and up to 1,000 or more feet long. The banks of glaciers are often flanked by deep crevasses, where they meet the walls of the glacial valley. Crevasses also run parallel to each other down the entire length of the ice streams, especially when the central portion of the glacier flows faster than the outer edges.

The ice in East Antarctica rests on solid bedrock and is therefore firmly anchored on land, making it reasonably stable. In contrast, the ice in West Antarctica rests below the sea on the continental shelf and glacial till. It is surrounded by floating ice that is pinned in by small islands buried below. Glacial mapping and radar surveys have suggested that the West Antarctic ice sheet is inherently unstable. Most of the western landmass is hundreds of feet below sea level. As long as sea levels remain the same, the ice continues to rest on the bottom. However, if sea levels rise, the ice sheet could break free of its moorings and began to float.

A warmer climate could cause the West Antarctic ice sheet to collapse suddenly. The rapidly melting, unstable ice sheet would break loose and crash into the sea. The additional sea ice would raise global sea levels upward of 20 feet and inundate coastal areas, a hazard scientists take very seriously. Even a slow melting of both polar ice caps would raise the level of the ocean upward

Figure 145 *Crevasses in Sperry Glacier partially filled with snow, Glacier National Park, Flathead County, Montana, on August 16, 1913.*

(Photo by W. C. Alden, courtesy USGS)

of 12 feet by the end of this century, drown much of the world's coastal plains, and flood coastal cities. A rise in sea level would also lift West Antarctic ice shelves off the seafloor and set them adrift into warm equatorial waters, where they would rapidly melt and raise the sea higher still.

GLACIAL SURGE

About 200 surge glaciers exist in North America. Some of them seem destined to cross the Alaskan oil pipeline that brings petroleum from the North Slope to the seaport of Valdez. During most of their lives, surge glaciers behave as normal glaciers do, traveling at a snail's pace of perhaps a couple of inches per day. However, at intervals of 10 to 100 years, the glaciers gallop forward up to 100 times faster than usual. One dramatic example is the Bruarjokull Glacier in Iceland. In a single year, it advanced 5 miles, at times achieving speeds of 16 feet an hour.

The surge often progresses along a glacier like a great wave, proceeding from one section to another with a motion similar to that of a caterpillar. Subglacial streams of meltwater act as a lubricant, allowing the glacier to flow rapidly toward the sea. Increasing water pressure under the glacier can lift it off its bed, overcoming the friction between ice and rock. This frees the glacier, which quickly slides downhill under the pull of gravity. The reason for glacial surge has not yet been fully explained, although they might be influenced by the climate, volcanic heat, and earthquakes. However, surge glaciers exist in the same regions as normal glaciers, often almost side by side. Additionally, for unexplained reasons, the great 1964 Alaskan earthquake failed to produce more surges than usual (Fig. 146).

Ice cliffs of the Columbia Glacier, located just west of Valdez, towered 300 feet above the deep water of the Prince William Sound. For hundreds of years, the glacier has been calving icebergs and smaller chunks of ice called growlers, bergy bits, and brash ice that often make shipping difficult in the sound. During the early 1980s, the glacier was in full retreat, moving rapidly backward more than 1.5 miles in 4 years. Many other glaciers have been retreating over the last 100 to 150 years.

The Bering Glacier in southern Alaska is the largest and longest surge glacier in North America and the biggest temperate surging glacier on Earth. In June 1993, the glacier began speeding downslope up to 300 feet a day for the first time in 26 years. The surge began in the middle of the glacier and expanded until it reached the seaward end. The 120-mile-long glacier has experienced similar phases of rapid movement with up to 2 years duration around 1900, 1920, 1940, and the late 1950s, with the last one ending in 1967.

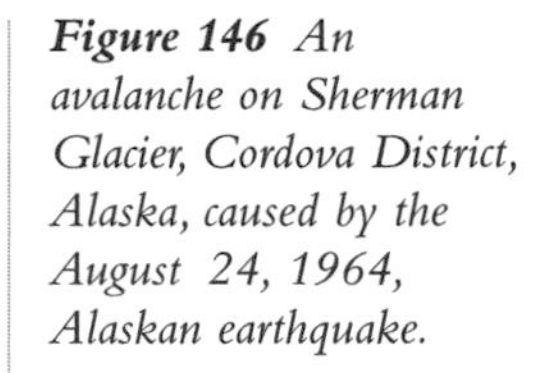

Figure 146 *An avalanche on Sherman Glacier, Cordova District, Alaska, caused by the August 24, 1964, Alaskan earthquake.*

(Photo by A. Post, courtesy USGS)

Normally, the glacier moves at a rate of about 10 feet per day. However, this last surge accelerated this river of ice up to many times greater than normal speed. In just three weeks, the glacier's terminus moved forward about a mile. As the glacier continued to surge, it displayed a variety of features, including deeply crevassed bulges and pressure ridges, extensional fractures with intricate patterns of intersecting crevasses, tear faults, temporary lakes, grabens, and many other stress-generated features. In July 1994, an outburst flood began from under the face of the margin of the glacier. As a result, a large discharge of sediment-laden water and huge blocks of ice spouted forth from the glacier.

Figure 147 *The Hubbard Glacier, Russell Fiord, Yakutat District, Alaska Gulf region, Alaska.*

(Photo by Austin Post, courtesy USGS)

Some 800 years ago, Alaska's Hubbard Glacier (Fig. 147) charged toward the sea. It later retreated. Then it advanced again 500 years later. Since 1895, the 70-mile-long river of ice has been flowing steadily toward the Gulf of Alaska at a rate of about 200 feet per year. In June 1986, however, the glacier surged ahead as much as 47 feet a day. Meanwhile, a western tributary, called Valerie Glacier, advanced up to 112 feet per day. Hubbard's surge closed off Russell Fjord with a formidable ice dam, some 2,500 feet long and up to 800 feet high, whose caged waters threatened the town of Yakutat to the south.

Some 20 similar glaciers around the Gulf of Alaska are heading toward the sea. If a significant number of these surge glaciers reach the ocean, they could significantly raise sea levels. With higher seas, West Antarctic ice shelves could rise off the seafloor and become adrift. A flood of ice would then surge into the Southern Sea. With the continued rise in sea level, more ice would plunge into the ocean, causing seas to rise even higher, which in turn would release more ice and set in motion a vicious circle.

The additional sea ice floating toward the Tropics would increase the Earth's albedo by reflecting sunlight back into space and lower global temperatures, perhaps enough to initiate glaciation. This scenario appears to have been played out at the end of the last warm interglacial, called the Sangamon.

During this period, sea ice cooled the ocean dramatically like ice cubes in a glass of water, spawning the beginning of the most recent Ice Age.

RISING SEA LEVELS

Sea levels have always been changing. More than 30 rises and falls of global sea levels occurred between 2 and 6 million years ago. At its highest point between 3 and 5 million years ago, the global sea level rose about 140 feet higher than it is today. Between 2 and 3 million years ago, the sea level dropped at least 65 feet lower than at present. During the Ice Age, sea levels dropped as much as 400 feet at the peak of glaciation. If all the ice on Antarctica melted, it would supply enough water to the ocean to raise global sea levels nearly 200 feet.

For centuries, civilizations have had to cope with changing sea levels (Table 9). If the seas continue to rise, the Dutch, who reclaimed their land from the sea, would find a major portion of their country lying under water. Many islands would drown or become mere skeletons of their former selves with only their mountainous backbones showing above the water. Most of the major cities of the world, because they are located on seacoasts or along inland waterways, would be inundated by the sea, with only the tallest skyscrapers poking above the waterline.

As global temperatures increase, coastal regions, where half the people of the world live, would feel the adverse effects of rising sea levels due to melting ice caps and thermal expansion of the ocean. Surface waters off the California coast have warmed nearly 1 degree Celsius over the past half century, causing the ocean to expand and raise the sea level about 1.5 inches. Higher sea levels are also due in part to the sinking of the land because of the increased weight of seawater pressing down onto the continental shelf. The additional freshwater in the North Atlantic could affect the flow of the Gulf Stream, causing Europe to freeze while the rest of the world swelters.

Over the last century, the global sea level appears to have risen upward of 9 inches due mainly to the melting of the Antarctic and Greenland ice sheets. The melting of the Arctic ice pack, like ice cubes dissolving in a cold drink, would not significantly raise the level of the ocean. The rapid deglaciation at the end of the last Ice Age between 16,000 and 6,000 years ago, when torrents of meltwater entered the ocean, raised the sea level on a yearly basis just 10 times greater than it is rising today. The present rate of sea level rise is several times faster than 40 years ago, amounting to about an inch every 5 years. Most of the increase appears to result from melting ice sheets.

Alpine glaciers, which contain substantial quantities of ice, appear to be melting as well (Fig. 148), possibly due to a warmer climate. Some areas such

TABLE 9 MAJOR CHANGES IN SEA LEVELS

Date	Sea Level	Historical Event
2200 B.C.	low	
1600 B.C.	high	Coastal forest in Britain inundated by the sea.
1400 B.C.	low	
1200 B.C.	high	Egyptian ruler Ramses II builds first Suez Canal.
500 B.C.	low	Many Greek and Phoenician ports built around this time are now under water.
200 B.C.	normal	
A.D. 100	high	Port constructed well inland of present-day Haifa, Israel.
A.D. 200	normal	
A.D. 400	high	
A.D. 600	low	Port of Ravenna, Italy, becomes landlocked. Venice is built and is presently being inundated by the Adriatic Sea.
A.D. 800	high	
A.D. 1200	low	Europeans exploit low-lying salt marshes.
A.D. 1400	high	Extensive flooding in low countries along the North Sea. The Dutch begin building dikes.

as the European Alps appear to have lost more than half their cover of ice. Moreover, the rate of loss apparently is accelerating. Tropical glaciers such as those in Indonesia have receded at a rate of 150 feet per year over the last two decades.

If the present melting continues, the sea could rise a foot or more by the year 2030. For every foot of sea level rise, 100 to 1,000 feet of shoreline will be inundated, depending on the slope of the coastline. Just a 3-foot rise could flood about 7,000 square miles of coastal land in the United States, including most of the Mississippi Delta, possibly reaching the outskirts of New Orleans. The receding shoreline would result in the loss of large tracks of coastal land along with shallow barrier islands that protect the coast from storms. Low-lying fertile deltas that support millions of people would disappear. Delicate wetlands, where many species of marine life hatch their young, would also vanish.

Vulnerable coastal cities would have to rebuild farther inland or construct protective seawalls to hold back the ocean. Other parts of the world would fare much worse. Half the scattered islands of the Republic of Maldives southwest of India would be lost. Much of Bangladesh also would drown, a

Figure 148 *White Chuck Glacier at Glacier Peak Wilderness Area, Snohomish County, Washington, began retreating in 1949.*

(Photo by A. Post, courtesy USGS)

particularly distressing situation since the people there can barely support themselves off the land.

Steep waves that accompany storms at sea cause serious beach erosion in the United States (Fig. 149). The constant pounding of the surf also erodes most human-made defenses against the rising sea. Upward of 90 percent of America's once sandy beaches are sinking beneath the waves. Barrier islands and sandbars running along the East Coast and East Texas are disappearing at alarming rates. Beaches along North Carolina are retreating at a rate of 4 to 5 feet per year. Sea cliffs are eroding back several feet a year, often destroying expensive homes. Most defenses used in a fruitless attempt to stop beach erosion usually end in defeat as waves relentlessly batter the coast.

The sea level is rising up to 10 times faster than it did a century ago, amounting to about a quarter of an inch per year. Most of the increase appears to result from melting ice caps, particularly in West Antarctica and Greenland. Greenland holds about 6 percent of the world's freshwater in its ice sheet. An apparent warming climate is melting more than 50 billion tons of water a year from the Greenland ice sheet, amounting to over 11 cubic miles of ice annually. The melting of the Greenland ice and the calving of icebergs from glaci-

Figure 149 *Property damage during a storm at Kitty Hawk, North Carolina.*

(Photo by R. Dolan, courtesy USGS)

ers entering the sea is responsible for about 7 percent of the yearly rise in global sea level. The Greenland ice sheet is undergoing significant thinning of the southern and southeastern margins, in places as much as 7 feet a year. Furthermore, Greenland glaciers are moving more rapidly to the sea, possibly caused by meltwater at the base of the glaciers that helps lubricate the downhill slide of the ice streams. The melting is increasing the risk of coastal flooding around the world during high tides and storms.

More icebergs are calving off glaciers entering the sea. They appear to be getting larger as well, threatening the stability of the ice sheets. Much of this instability is blamed on global warming. To determine whether greenhouse gases are actually causing the planet to warm, scientists are studying the speed at which sound waves travel through the ocean. Since sound travels faster in warm water than in cold water—a phenomenon known as acoustic thermometry—long-term measurements could reveal whether global warming is a certainty. The idea is to send out low-frequency sound waves from a single station and monitor them from several listening posts scattered around the world. The signals take several hours to reach the most distant stations. Therefore, shaving a few seconds off the travel time over an extended period of 5 to 10 years could definitely indicate that the oceans are indeed warming.

The first direct measurements of a possible ocean warming from rising global temperatures were taken by satellites. The polar sea ice appears to have shrunk by as much as 6 percent since the 1970s when measurements were first made. However, an extensive study of temperatures over the Arctic Ocean indicates the region has not warmed significantly over the last four decades. Perhaps the Arctic is the last region affected by greenhouse warming. Sea ice covers most of the Arctic Ocean and forms a frozen band around Antarctica during the winter season in each hemisphere. If global warming melts the polar sea ice, the number of microscopic organisms could dramatically fall as well as the marine animals that feed on them, adversely affecting the ecology of the world's oceans.

During the Sangamon interglacial prior to the last Ice Age, the melting of the ice caps caused the sea level to rise about 60 feet higher than at present. If average global temperatures continue to rise, this interglacial could become equally as warm if not warmer than the last one. The warmer climate could induce an instability in the West Antarctic ice sheet, causing it to surge into the sea. This rapid flow of ice into the ocean could raise sea levels up to 20 feet, inundate the continents as much as 3 miles inland, and flood trillions of dollars of property.

After discussing the role of glacial ice in geology, including landform development, their effect on climate, and influence on sea levels, the next chapter searches for asteroid and comet impacts over the Earth, why they occur, where they are likely to fall, and their effects on life.

9

IMPACT CRATERING

COSMIC INVASION OF THE EARTH

This chapter examines the most destructive of geologic forces responsible for raining havoc down onto the Earth. The ultimate environmental hazard is the mass destruction caused by asteroid or comet impacts on the Earth. More than 150 large meteorite craters have been identified throughout the world (Fig. 150), some of which might be associated with mass extinctions of species.

Sometimes asteroids the size of mountains struck the planet, inflicting a great deal of damage and causing tragic extinction episodes. Massive comet swarms, involving perhaps thousands of comets impacting all over the Earth, might also explain the disappearance of species. Occasionally, a wayward asteroid wanders near the Earth. Should one ever impact the surface, it would create as much havoc as a global nuclear war. Indeed, the environmental consequences would be similar to those of a hypothetical nuclear winter, which would make survival difficult for all creatures.

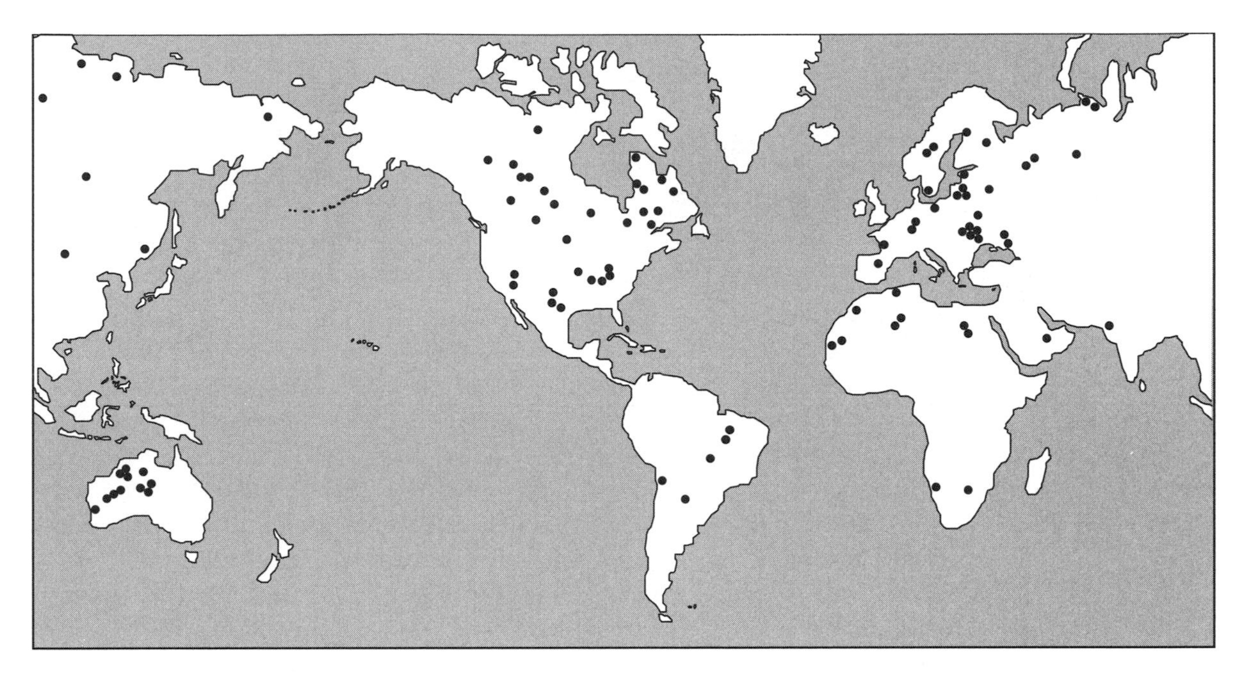

Figure 150 The locations of some known major meteorite impact structures around the world.

THE ASTEROID BELT

Asteroids, from Greek meaning "starlike," are leftovers from the creation of the solar system. The asteroid belt lies between the orbits of Mars and Jupiter and comprises about a million pieces of solar system rubble larger than a mile across along with more numerous smaller objects. Not all asteroids reside within the main belt, however. An interesting group called Trojans lies in the same orbit as Jupiter. Another asteroidlike body has a far-ranging orbit that carries it from near Mars to beyond Uranus. One large object called Chiron lies between the orbits of Saturn and Uranus, indeed an odd place for an asteroid.

Asteroids are a relatively recent discovery. On January 1, 1801, while searching for the "missing planet" in the wide region between Mars and Jupiter, the Italian astronomer Giuseppe Piazzi instead discovered the asteroid Ceres, named for the guardian goddess of Sicily. It is the largest of the known asteroids, with a diameter of more than 600 miles.

Because of the strong gravitational attraction of Jupiter, a ring of planetesimals surrounding the Sun was unable to coalesce into a single planet. Instead, these primordial fragments formed several planetoids smaller than the Moon as well as a broad band of debris, called meteoroids. These are fragments broken off asteroids by numerous collisions. Originally, the combined masses

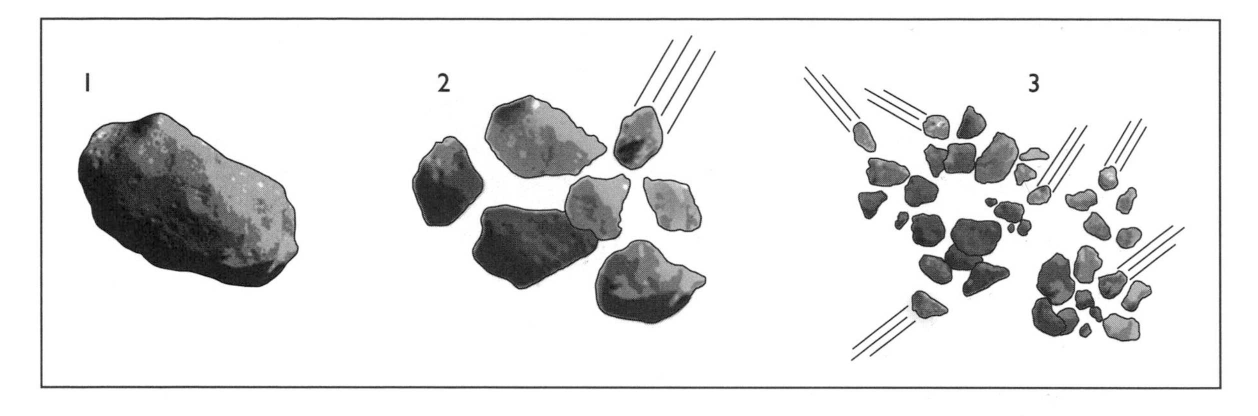

Figure 151 *A planetoid less than the size of the Moon is broken up by a collision with another body, and additional collisions yield asteroids that bombard the Earth.*

of all the material in the asteroid belt were nearly equal to the present mass of the Earth. However, constant collisions have weeded out the asteroids, so now their combined mass is perhaps less than 1 percent of the original.

Many asteroids contain a high concentration of iron and nickel, suggesting they were once part of the metallic core of a planetoid that disintegrated after a collision with another body. Some large asteroids might have melted and differentiated early in the formation of the solar system. The inner- and middle-belt asteroids underwent a great deal of heating and experienced as much melting as did the planets. The molten metal in the asteroids along with siderophiles (iron lovers), such as iridium and osmium of the platinum group, sank to their interiors and solidified. The metallic cores were exposed after eons of asteroid collisions chipped away the more fragile surface rock. Final breakup after collisions between asteroids yielded several dense, solid fragments (Fig. 151).

Stony asteroids are much less dense and contain a high percentage of silica. They exist near the inner part of the asteroid belt. The darker carbonaceous asteroids, which contain a high percentage of carbon, lie toward the outer portion of the asteroid belt. Between these regions are wide spaces called Kirkwood gaps, named for American mathematician Daniel Kirkwood, that are almost totally devoid of asteroids. If an asteroid falls into one of these gaps, its orbit stretches, causing it to swing in and out of the asteroid belt. This brings it close to the Sun and the orbits of the inner planets, including the Earth.

CRATERING EVENTS

Between 4.2 and 3.8 billion years ago, thousands of large impactors the size of asteroids bombarded the Earth and the Moon (Fig. 152). All the inner planets and the moons of the outer planets display numerous craters from this massive meteorite shower. Fortunately, little large-scale asteroid activity has happened since. The bombardment melted the Earth's thin basaltic crust by impact friction. The meteorites blasted apart the crust to form huge impact basins, some with walls nearly 2 miles above the surrounding terrain and floors 10 miles deep.

During the height of the great meteorite bombardment, an enormous asteroid slammed into the North American continent in what is now central

Figure 152 *The lunar terrain, showing numerous large meteorite craters and expansive lava plains.*

(Photo courtesy NOAO)

Ontario, Canada, possibly creating a crater upward of 900 miles wide. The Earth of that time was covered by a global ocean. The giant impact might have triggered the formation of continents. About 1.8 billion years ago, Ontario was struck again by a large meteorite, generating enough energy to melt vast quantities of rock. The impact created the Sudbury Igneous Complex, the world's largest and richest nickel ore deposit. This is also the site of one of the world's oldest eroded impact structures, called astroblemes. Scattered about the location are shatter cones, which are striated, conically shaped rocks fractured by shock waves generated by large meteorite impacts.

More than a mile below the floor of Lake Huron lies a 30-mile-wide rimmed circular remnant of an apparent impact structure produced by a large meteorite some 500 million years ago. A crater this size would have required the impact of a meteorite about 3 miles wide. It is only one of many large impact craters scattered around the world during the last half billion years. Roughly 365 million years ago, two distinct meteorite or comet impacts on the Asian continent appear to have caused major extinctions in the late Devonian period.

A large impact at the end of the Triassic period 210 million years ago created the Manicouagan structure in Quebec, Canada (Fig. 153). The Manicouagan River and its tributaries form a reservoir around a roughly circular structure some 60 miles across, making it one of the six largest craters known on Earth. An almost perfectly circular ring of water, produced when sections of the crater flooded, surrounds the raised center of the impact structure. The structure is composed of Precambrian rocks reworked by shock metamorphism generated by the impact of a large celestial body.

The Saint Martin impact structure northwest of Winnipeg, Manitoba, is 25 miles wide and mostly hidden beneath younger rocks. Three other impact structures include the 16-mile-wide Rochechouart in France, a 9-mile-wide crater in Ukraine, and a 6-mile-wide crater in North Dakota. The craters appear to have formed about the same time roughly 210 million years ago. This date coincides with a mass extinction at the end of the Triassic that eliminated nearly half the reptilian families and paved the way for the dinosaurs, which ruled the Earth for the next 145 million years.

One of the world's largest impact craters covers most of western Czech Republic centered near the capital city Prague. It is about 200 miles in diameter and at least 100 million years old. Concentric circular elevations and depressions surround the city, which is what would be expected if the Prague Basin were indeed a meteorite crater. Moreover, green tektites (small, rounded, glasslike stones) created by the melt from an impact were found in an arc that follows the southern rim of the basin. The circular outline was discovered in a weather satellite image of Europe and North Africa, and its immense size probably kept it from being noticed earlier.

Figure 153 *The Manicouagan impact structure, Quebec, Canada.*

(Courtesy NASA)

About 65 million years ago, another large meteorite supposedly struck the Earth, creating a crater at least 100 miles wide. The debris sent the planet into environmental chaos. Meteorite fallout materials lie atop 65-million-year-old sediments worldwide. The search for the meteorite impact site has been concentrated around the Caribbean area (Fig. 154). There thick deposits of wave-deposited rubble have been found along with melted and crushed rock ejected from the crater. A large asteroid appears to have struck near the present town of Chicxulub, on the Yucatán Peninsula, Mexico, creating the explosive force of 100 trillion tons of TNT, or 1,000 times more powerful than the detonation of all the world's nuclear arsenals.

The crater is 110 miles wide, among the largest known impact features on Earth. It lies beneath 600 feet of sedimentary rock on the northern coast of the Yucatán Peninsula. If the meteorite landed on the seabed just offshore, 65 million years of sedimentation would have long since buried it under thick

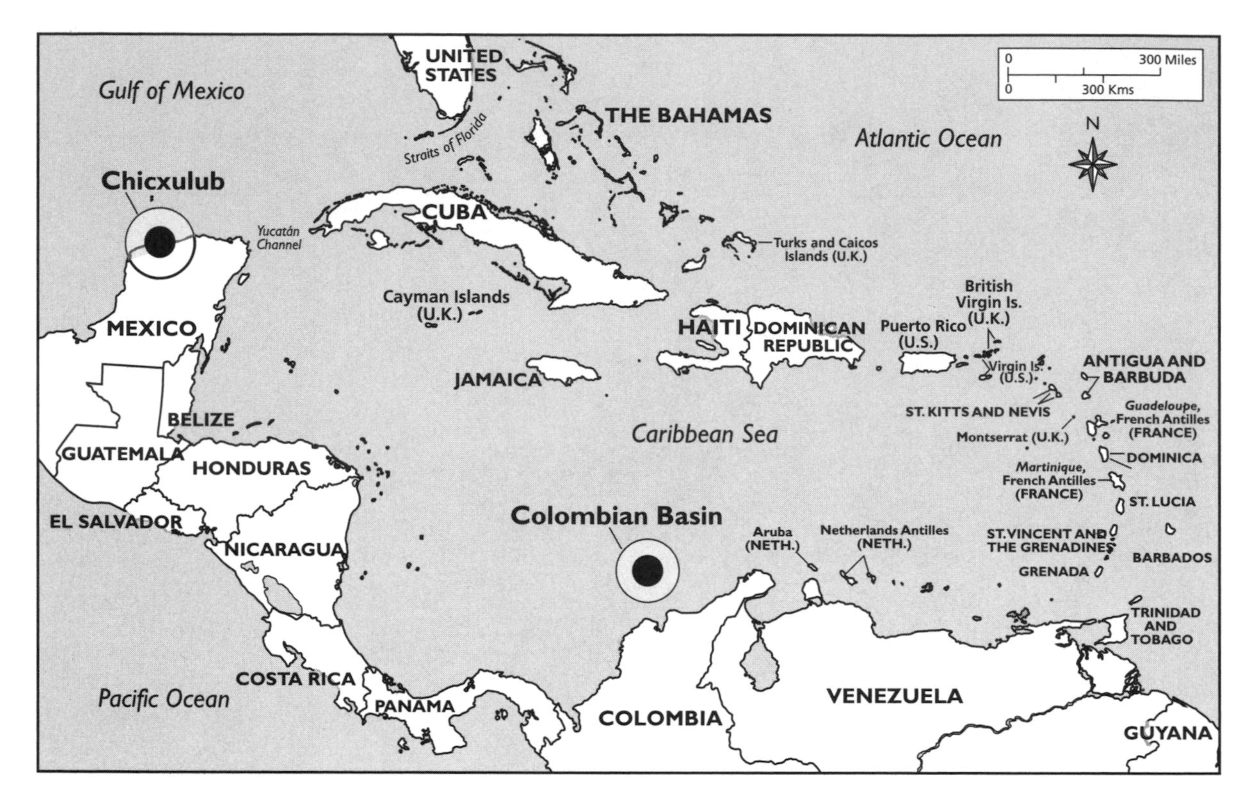

Figure 154 *Possible impact structures in the Caribbean area that might have ended the Cretaceous period.*

deposits of sand and mud. Furthermore, a splashdown in the ocean would have created an enormous tsunami that would have scoured the seafloor and deposited its rubble onto nearby shores. The impact is thought to have been responsible for the extinction of the dinosaurs along with more than half of all other species, mostly terrestrial animals and plants. Therefore, the dinosaurs might have been both created and destroyed by large meteorite impacts.

The most pronounced undersea impact crater is the 35-mile-wide Montagnais structure 125 miles off the southeast coast of Nova Scotia (Fig. 155). The crater is 50 million years old and closely resembles craters on dry land, except its rim is 375 feet beneath the sea and the crater bottom is 9,000 feet deep. The crater was created by a large meteorite up to 2 miles wide. The impact raised a central peak similar to those seen inside craters on the Moon. The structure also contained rocks melted by a sudden shock. Such an impact would have sent a tremendous tsunami crashing down onto nearby shores.

Some 40 million years ago, perhaps two or three large meteorite impacts possibly caused another mass extinction, killing off the archaic mammals. These were large, grotesque creatures whose disappearance paved the way for the evolution of modern mammals. In addition, major European mountain ranges were forming during this time. The upward thrust of so much crust might have substantially cooled the planet, killing large numbers of species that could not adapt to the cold.

About 23 million years ago, an asteroid or comet slammed into Devon Island in the Canadian Arctic with such a force that rock from more than half a mile underground shot skyward. The meteorite gouged out a 15-mile-wide hole called the Haughton Crater. Pulverized granite gneiss blown upward by the impact fell back to earth as hot breccia. Plant and animal life ceased to exist within a radius of perhaps 100 miles. At that time, the area was much warmer and lusher, with spruce and pine forests. Today, the crater is used as a testbed for future excursions to Mars, which has strikingly similar craters.

About 50,000 years ago, a large meteorite landed in northern Arizona near the present town of Winslow. The impact ejected nearly 200 million tons of rock and excavated a crater 4,000 feet wide and 560 feet deep, with the crater rim rising 135 feet above the desert floor. The impactor released the equivalent energy of about 20 megatons of TNT, equal to the most powerful nuclear weapons. The massive explosion pulverized rock material and deposited it around the crater to a maximum depth of 75 feet. Scattered out-

Figure 155 Location of the Manicouagan and Montagnais impact structures in North America.

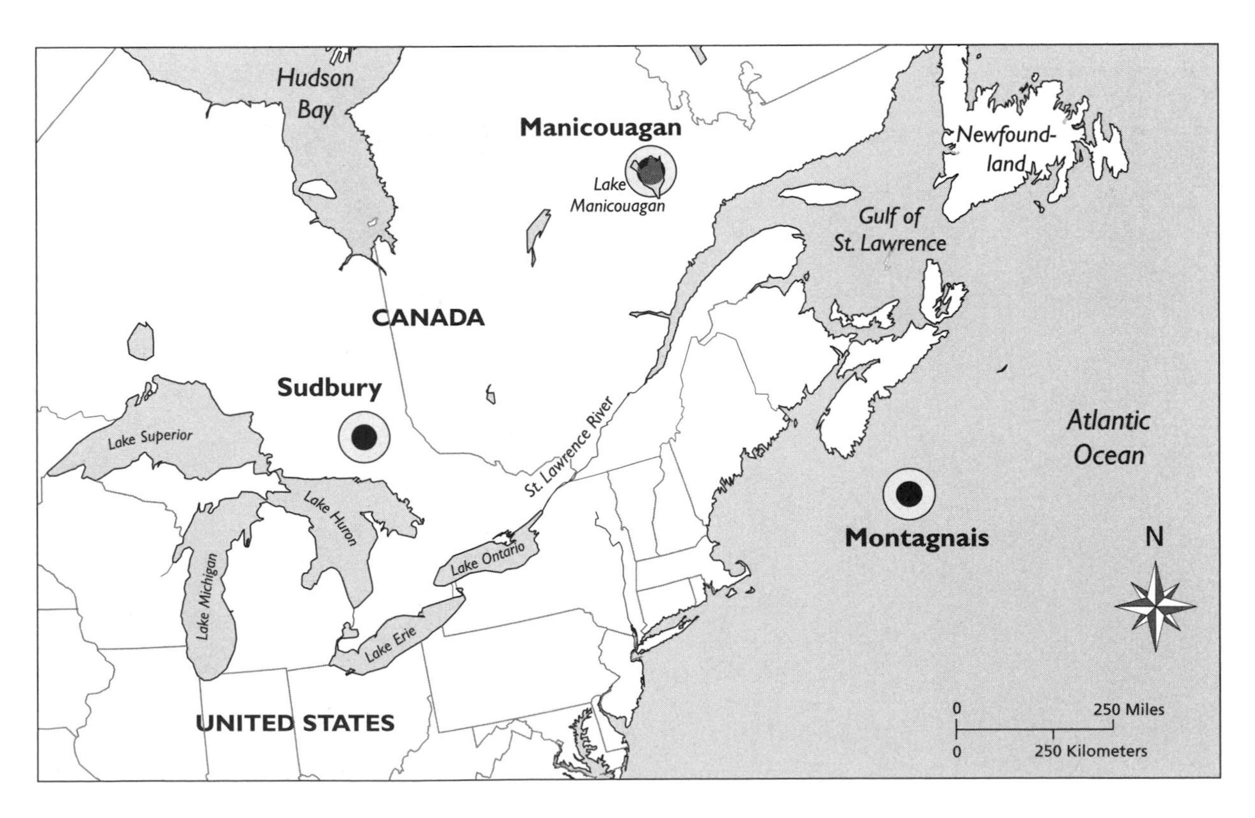

ward from the crater were several tons of metallic meteoritic debris. This indicates the meteorite was the iron-nickel variety, measuring about 200 feet in diameter and weighing about 1 million tons. Today, the crater is a tourist attraction called Meteor Crater (Fig. 156), also known as Barringer Crater, where people pay to view one of the best-preserved craters in the world. Craters formed in desert or tundra regions are better preserved because little erosion occurs in these areas.

The New Quebec Crater in Quebec, Canada, is the largest known meteorite impact structure where actual meteoritic debris has been found. It is a relatively young crater, perhaps only a few thousand years old, with a diameter of about 11,000 feet and a depth of about 1,300 feet. The crater contains a deep lake, whose surface lies 500 feet below the crater rim.

As little as 3,000 years ago, a meteorite appears to have blasted a mile-wide crater 12 miles west of Broken Bow, Nebraska. Another young impact structure, called the Wolf Creek Crater, lies near Halls Creek in Western Australia. Scattered around the crater were large pieces of the original meteorite (Fig. 157). The crater is rather shallow, with a diameter of 2,800 feet and a depth of 140 feet. Its presence is a constant reminder that large meteorite impacts are an ongoing process and another collision could occur anytime.

Figure 156 *Meteor Crater near Winslow, Arizona.*

(Courtesy USGS)

Figure 157 *Crack development in the Wolf Creek meteorite from Western Australia.*

(Photo by G. T. Faust, courtesy USGS)

CRATERING RATES

The heavily cratered lunar highland is the most ancient region on the Moon and contains a record of intense bombardment from around 4 billion years ago. Generally, the older the surface, the more craters it comprises. In time, the number of impacts rapidly declines, and the impact rate remains low due to a depletion of asteroids and comets. The rate of cratering appears to differ from one part of the solar system to another. The cratering rates and the total number of craters suggest that the average rates over the past few billion years were similar for the Earth, the Moon, and the rest of the inner planets.

Generally, the cratering rates for the moons of the outer planets appear to be significantly lower than those for the inner solar system, where most of the asteroids lie. Nonetheless, the crater sizes are comparable to those on the Earth's moon and Mars. Cratering rates for the Moon and Mars appear to have been nearly equal. However, because Mars is nearer the asteroid belt, the impact rates for Mars were probably higher than those for the Moon.

Major obliteration events have occurred on Mars as recently as 200 to 400 million years ago, whereas most of the scarred lunar terrain is billions of years old. Mars has erosional agents, such as wind and ice, that tend to erase

Figure 158 *A heavily cratered region on Mars showing the effects of wind erosion.*
(Courtesy NASA)

impact craters (Fig. 158). In contrast, the dominant mechanism for destroying craters on the Moon is other impacts. Moreover, the high degree of crater overlap makes placing the craters into their proper geologic order difficult.

On Earth, impact craters range from a few thousand to nearly 2 billion years old. Over the past 3 billion years, the cratering rate has been fairly constant, with a major impact producing a crater 30 miles or more in diameter occurring every 50 to 100 million years. As many as three large meteorite impacts that produce craters about 10 miles wide occur every million years. An asteroid half a mile in diameter, with an impact energy of a million megatons of TNT, capable of wiping out a quarter of the world's population, could strike the Earth every 100,000 years or so. A meteorite several hundred feet wide impacts every 200 to 300 years, producing the energy equivalent of a multimegaton nuclear weapon that could level a large city.

Of all the known impact craters distributed around the world, most are younger than 200 million years. The older craters are less plentiful because erosion and sedimentation have destroyed them. Consequently, only 10 percent of all large craters less than 100 million years old have been discovered. Most of the known impact craters exist in stable regions in the interiors of continents because these areas experience low rates of erosion and other destructive processes.

METEORITE IMPACTS

The Earth's highly active geology has long since erased all but the faintest signs of ancient impact craters. Impacts on other bodies in the solar system are quite evident and numerous (Fig. 159). However, the Earth appears as though it escaped such a heavy bombardment. Nonetheless, because of its larger size and

Figure 159 *The heavily cratered terrain on Mercury.*

(Courtesy NASA)

greater gravitational attraction, the Earth endured several times more meteorite impacts than its next-door neighbor, the Moon. Therefore, our planet was just as heavily bombarded as the rest of the solar system, but only vague remnants of ancient craters remain. Many circular features appear to be impact structures. However, due to their low profiles and subtle stratigraphies, they previously went unrecognized as impact craters.

Many geologic features once believed to be formed by forces such as uplift are now thought to be impact craters. For example, Upheaval Dome near the confluence of the Colorado and Green Rivers in Canyon Lands National Park, Utah, was originally thought to be a salt plug that heaved the overlying strata upward into a huge bubblelike fold 3 miles wide and 1,500 feet high. However, an alternative interpretation holds that the structure is actually a deeply eroded astrobleme, which is the remnant of an ancient impact structure gouged out by a large cosmic body striking the Earth such as an asteroid or comet.

Erosion has removed as much as a mile or more of the overlying beds since the meteorite smashed into the ground between 30 and 100 million years ago, making the structure possibly the planet's most deeply eroded impact crater. The original crater apparently made a 4.5-mile-wide hole in the ground that has been heavily modified by deep erosion over the many years. The dome itself appears to be a central rebound peak similar to those on the Moon formed when the ground heaved upward by the force of the impact. The meteorite measured about 1,700 feet wide and crashed to Earth with a velocity of several thousand miles per hour. On impact, it created a huge fireball that would have incinerated everything for hundreds of miles.

Some meteorite impacts result from multiple hits, leaving a chain of two or more craters close together. They are often produced when an asteroid or comet breaks up in outer space or upon entering the atmosphere. A rapid-fire impact from a broken-up object a mile wide apparently produced a string of three 7.5-mile-wide impact craters in the Sahara Desert of northern Chad. Two sets of twin craters, Kara and Ust-Kara in Russia and Gusev and Kamensk near the northern shore of the Black Sea, formed simultaneously only a few tens of miles apart. Splayed across southern Illinois, Missouri, and eastern Kansas are eight large, gently sloping depressions, 2 to 10 miles wide and averaging 60 miles apart. A chain of 10 oblong craters ranging up to 2.5 miles long and a mile wide, running along a 30-mile line near Rio Cuarto, Argentina, suggests that a meteorite 500 feet wide hit at a shallow angle and broke into pieces that ricocheted and gouged their way across the landscape roughly 2,000 years ago.

Dispersed throughout the world are a number of impact structures (Table 10). These are large, circular features created by the sudden shock of a large meteorite impacting onto the surface. The structures are generally circular or slightly oval in shape and range in size from 1 to 50 miles or more wide. Some meteorite impacts form lasting distinctive craters, while others

TABLE 10 LOCATION OF MAJOR METEORITE CRATERS AND IMPACT STRUCTURES

Name	Location	Diameter (in feet)
Al Umchaimin	Iraq	10,500
Amak	Aleutian Islands	200
Amguid	Sahara Desert	
Aouelloul	Western Sahara Desert	825
Baghdad	Iraq	650
Boxhole	Central Australia	500
Brent	Ontario, Canada	12,000
Campo del Cielo	Argentina	200
Chubb	Ungava, Canada	11,000
Crooked Creek	Missouri, USA	
Dalgaranga	Western Australia	250
Deep Bay	Saskatchewan, Canada	45,000
Dzioua	Sahara Desert	
Duckwater	Nevada, USA	250
Flynn Creek	Tennessee, USA	10,000
Gulf of St. Lawrence	Canada	
Hagensfjord	Greenland	
Haviland	Kansas, USA	60
Henbury	Central Australia	650
Holleford	Ontario, Canada	8,000
Kaalijarv	Estonia, USSR	300
Kentland Dome	Indiana, USA	3,000
Kofels	Austria	13,000
Lake Bosumtwi	Ghana	33,000
Manicouagan Reservoir	Quebec, Canada	200,000
Merewether	Labrador, Canada	500
Meteor Crater	Arizona, USA	4,000
Montagne Noire	France	
Mount Doreen	Central Australia	2,000
Murgab	Tadjikistan, USSR	250
New Quebec	Quebec, Canada	11,0000

(continues)

TABLE 10 (CONTINUED)

Name	Location	Diameter (in feet)
Nordlinger Ries	Germany	82,500
Odessa	Texas, USA	500
Pretoria Saltpan	South Africa	3,000
Serpent Mound	Ohio, USA	21,000
Sierra Madera	Texas, USA	6,500
Sikhote-Alin	Sibera, USSR	100
Steinheim	Germany	8,250
Talemzane	Algeria	6,000
Tenoumer	Western Sahara Desert	6,000
Vredefort	South Africa	130,000
Wells Creek	Tennessee, USA	16,000
Wolf Creek	Western Australia	3,000

present only outlines of former craters. The only evidence of their existence might be a circular disturbed area that contains rocks altered by shock metamorphism, which requires the instantaneous application of high temperatures and pressures similar to those found deep in the Earth's interior.

The most easily recognizable shock effect is the formation of shatter cones. These are caused by the fracturing of rocks into conical and striated patterns. They form most readily in fine-grained rocks that have little internal structure, such as limestone and quartzite. Large meteorite impacts also produce shocked quartz grains with prominent striations across crystal faces (Fig. 160). Minerals such as quartz and feldspar develop these features when high-pressure shock waves exert shearing forces onto their crystals, producing parallel fracture planes called lamellae.

The high temperatures developed by the force of the impact also fuse sediment into small glassy spherules, which are tiny spherical bodies. Extensive deposits of 3.5-billion-year-old spherules in South Africa are more than a foot thick in places. Spherules of a similar age have also been found in Western Australia. Spherule layers up to 3 feet thick have been found in the Gulf of Mexico and are related to the 65-million-year-old Chicxulub impact structure off Yucatán, Mexico. The spherules resemble the glassy chondrules (rounded granules) in carbonaceous chondrites, which are carbon-rich meteorites, and in lunar soils. The discoveries suggest that massive meteorite bom-

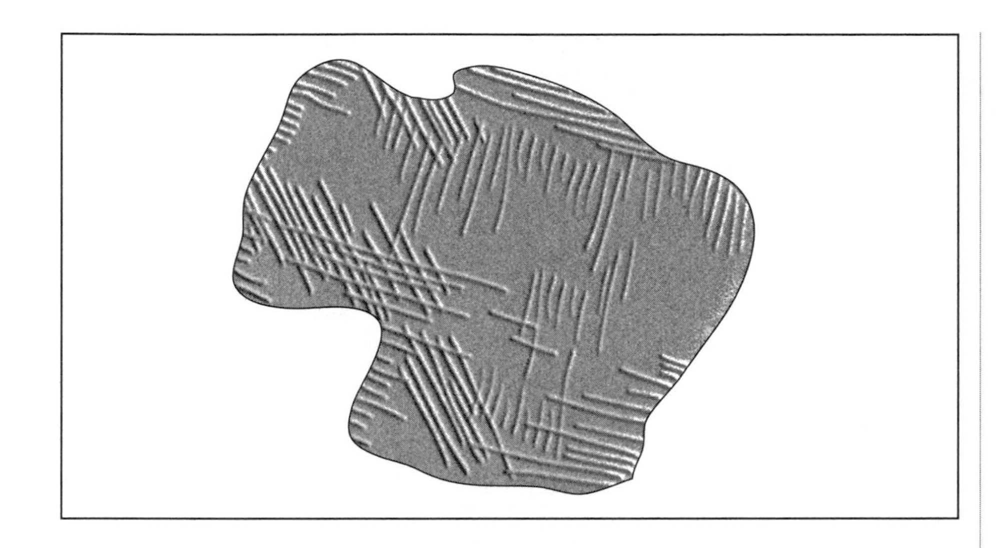

Figure 160 Lamellae across crystal faces produced by high-pressure shock waves from a large meteorite impact.

bardments during the early part of the Earth's history played a major role in shaping the surface of the planet.

When a meteorite slams into the Earth's surface, it generates a tremendous shockwave with a pressure of millions of atmospheres that travels down into the rock and reflects back up into the meteorite. As the meteorite burrows into the ground, it forces the rock aside, flattening itself in the process. It is then deflected and ejected from the crater, followed by a spray of shock-melted meteorite along with melted and vaporized rock that shoot out at tremendous velocities. The finer material lofts high into the atmosphere. Meanwhile, the coarse debris falls back around the perimeter of the crater, forming a high, steep-banked rim (Fig. 161).

Large meteorites traveling at high velocities completely disintegrate upon impact. In the process, they form craters generally 20 times wider than the meteorites themselves. The crater diameter also varies with the type of rock being impacted due to the relative differences in rock strength. A crater formed in crystalline rock such as granite might be twice as large as one made in sedimentary rock. Simple craters (Fig. 162) up to 2.5 miles in diameter form deep basins. In contrast, complex craters are much shallower, up to 100 times wider than they are deep. The craters generally contain a central peak surrounded by an annular trough and a fractured rim.

The Earth's highly active erosional processes have erased the vast majority of ancient impact structures. The exceptions are very large craters greater than 12 miles wide and more than 2.5 miles deep. These craters are so deep that even if erosion wears down the entire continent, faint remnants would still remain. Craters of extremely large meteorite impacts might temporarily reach depths of 20 miles or more and uncover the hot mantle below. The exposure

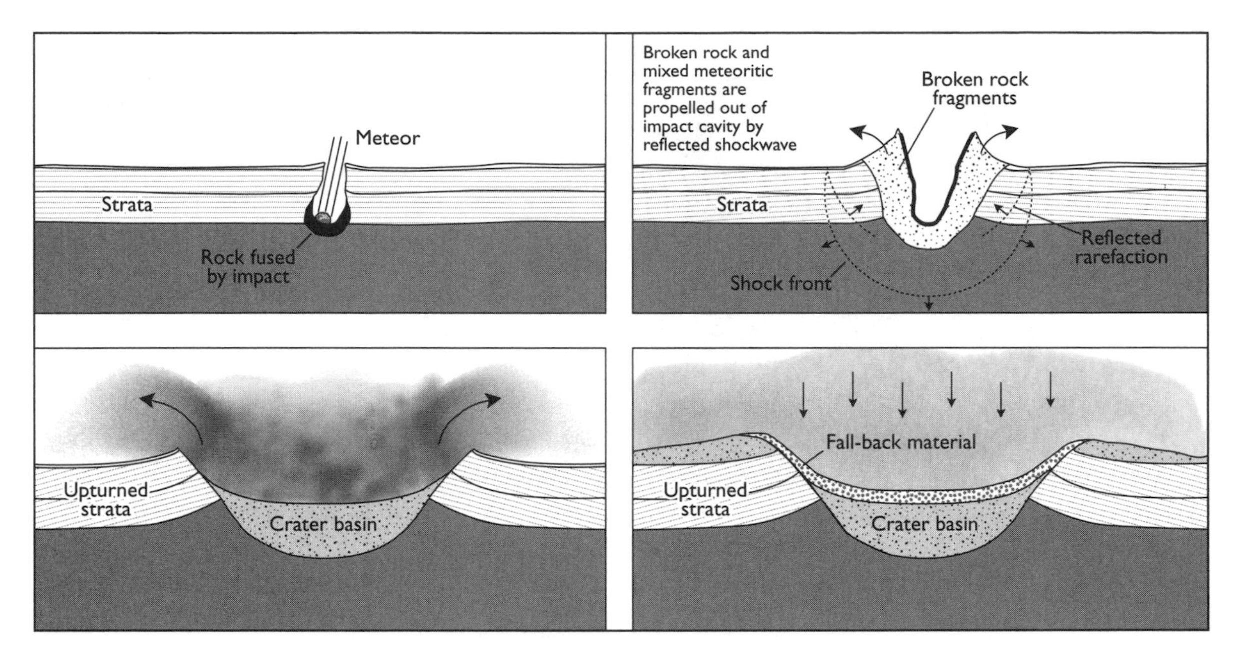

Figure 161 *The formation of a large meteorite crater.*

of the mantle in this manner would cause a gigantic volcanic explosion, releasing more material into the atmosphere than the meteorite impact itself.

ROGUE ASTEROIDS

Asteroids are minor planets, ranging up to hundreds of miles wide (Fig. 163). Most asteroids form a broad band of debris that orbits the Sun between Mars and Jupiter. The asteroid belt has an inclination of about 10 degrees with

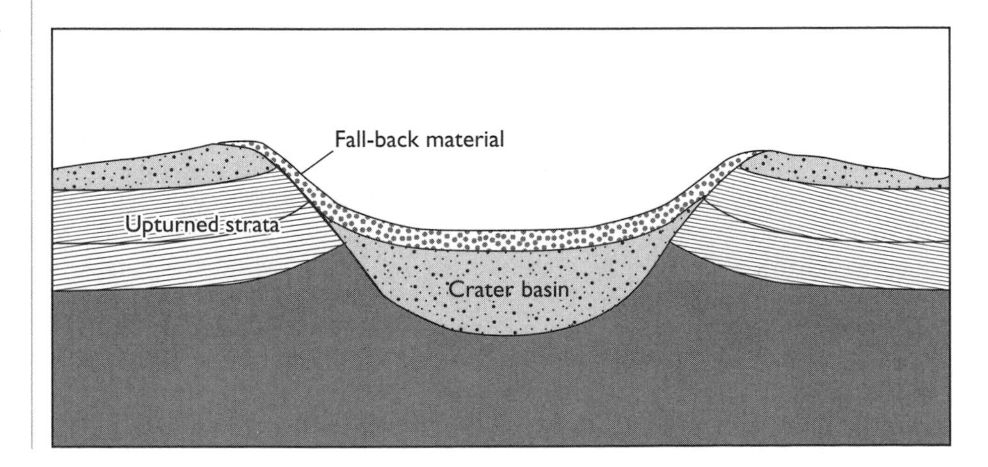

Figure 162 *Structure of a simple meteorite crater.*

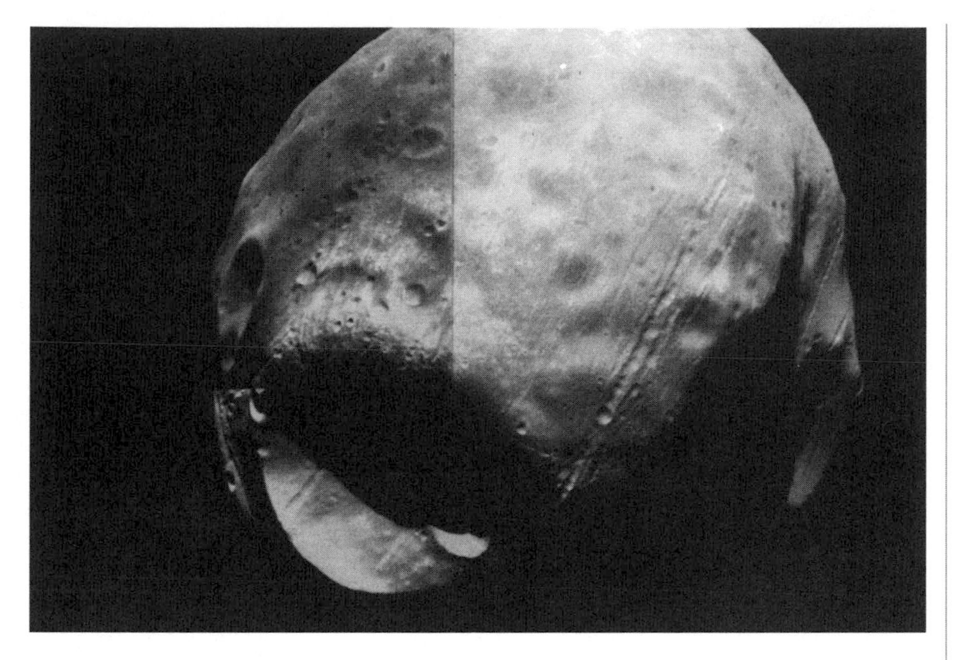

Figure 163 *Mars's inner moon, Phobos, might have been a captured asteroid 13 miles across.*
(Courtesy NASA)

respect to the ecliptic, the plane of the solar system. Of the million or so asteroids with diameters of half a mile or more, some 18,000 have thus far been located and identified. Of these, about 5,000 have had their orbits determined precisely. The orbits of the major asteroids have been accurately plotted so that space probes headed for the outer solar system can safely traverse the asteroid belt without sustaining a collision.

Most asteroids revolve around the Sun in elliptical orbits. Occasionally, some stretch out far enough to come within the orbits of the inner planets, including the Earth. About 60 asteroids have been observed to be out of the main asteroid belt and in Earth-crossing orbits. How they managed to fall into orbits that cross our planet's path remains a mystery. Apparently, the asteroids run in nearly circular orbits for a million or more years. For unknown reasons, their orbits then suddenly stretch and become so elliptical that they come within reach of our planet.

Some Earth-crossing asteroids, called Apollo asteroids, might have begun their lives outside the solar system and appear to be comets that have exhausted their volatile material composed of ices and gas after repeated encounters with the Sun. Dozens of Apollo asteroids have been identified out of a possible total of perhaps 1,000. Most are quite small and were discovered only when passing close by the Earth. Inevitable collisions with the Earth and other planets steadily depletes their numbers, requiring an ongoing source of Apollo asteroids, possibly derived from comet nuclei.

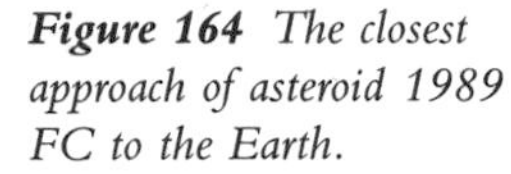

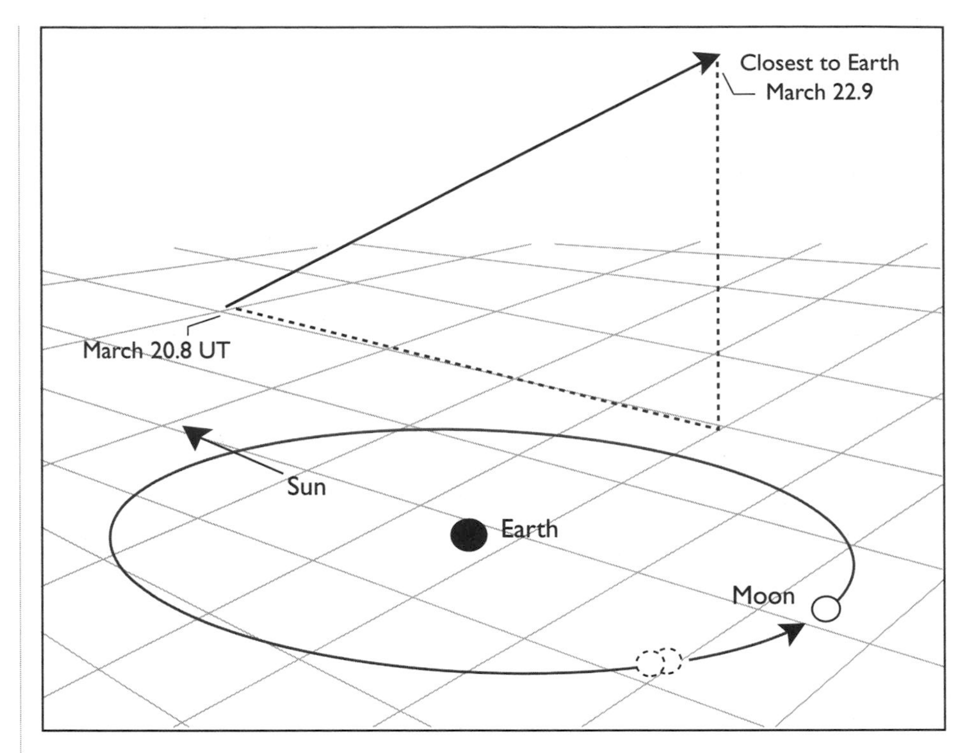

Figure 164 *The closest approach of asteroid 1989 FC to the Earth.*

One of the closest encounters with an asteroid occurred on October 30, 1937, when Hermes shot past the Earth at 22,000 miles per hour. The mile-wide asteroid missed hitting our planet by only half a million miles, or about twice the distance to the Moon. In astronomical terms, that was very close. If Hermes had collided with Earth, it would have released the energy equivalent of a 100,000-megaton hydrogen bomb. Indeed, nuclear war has many similarities to the impact of a large asteroid. The impact would send aloft huge amounts of dust and soot into the atmosphere. The debris would clog the skies and plunge the planet into a deep freeze for several months.

The closest flyby of a large asteroid in recorded history occurred on March 22, 1989, when asteroid 1989 FC came within 430,000 miles of our planet (Fig. 164). The asteroid was about half a mile wide. Though a collision with the Earth would have been catastrophic, a fluke of orbital geometry might have somewhat softened the blow. The asteroid orbits the Sun in the same direction as the Earth, completing a revolution in about one year and traveling at almost the same speed as this planet. Therefore, its approach was rather slow compared with other celestial objects. However, because of the Earth's large size, its gravitational pull would have accelerated the asteroid during its final approach. If a collision had occurred, the asteroid would have produced a crater 5 to 10 miles wide, large enough to wipe out a major city.

Astronomers did not detect asteroid 1989 FC until it was already moving away from the Earth. Only then did they notice a dramatic slowdown in the asteroid's motion against background stars. To their amazement, the asteroid was rushing straight away from Earth on what must have been a near-grazing trajectory. The astronomers failed to notice the asteroid's approach because it came from a sunward direction. Also, the Moon was nearly full, further hampering observations.

Asteroid 1989 FC is only one of 30 similar bodies that closely approach the Earth. In addition, several hundred to 1,000 or more asteroids wider than one-third mile are capable of crossing the Earth's orbit for a close encounter. For example, on December 8, 1992, a large asteroid called Toutatis, which measured 2.5 miles long by 1.6 miles wide, flew within 2.2 million miles of Earth. If it were only slightly off course and a collision had occurred, the effects could have threatened humanity.

Besides asteroids, comets have been known to fly nearby the Earth. Surrounding the Sun about a light-year away is a shell of more than a trillion comets with a combined mass of 25 Earths, called the Oort Cloud, named after the Dutch astronomer Jan Kendrick Oort. Another band of comets known as the Kuiper Belt exists closer to the Sun. However, it is still well beyond Pluto, which due to its odd orbit, is inclined 17 degrees to the ecliptic (plane of the solar system) and might itself be a captured comet nucleus or an asteroid.

Comets are hybrid planetary bodies consisting of a stony inner core and an icy outer layer (Fig. 165). They are characterized as flying icebergs mixed with small amounts of rock debris, dust, and organic matter. Comets are believed to be aggregates of tiny mineral fragments coated with organic compounds and ices enriched in the volatile elements hydrogen, carbon, nitrogen, oxygen, and sulfur. Comets might therefore be more accurately described as frozen mudballs with equal volumes of ice and rock.

Most comets travel around the Sun in highly elliptical orbits that carry them a thousand times farther out than the planets. Only when they swing

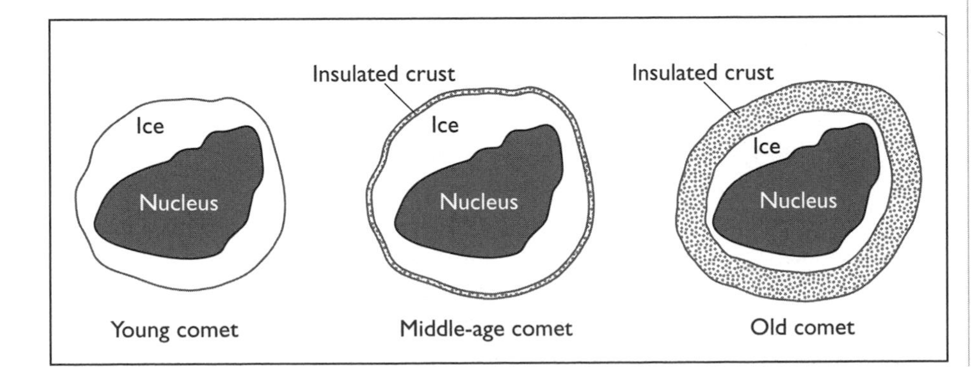

Figure 165 *Life cycle of a comet: 1—when it is young, fresh ice dominates the surface; 2—at middle age, the comet develops an insulated crust; and 3—at old age, the crust becomes so thick it cuts off all cometary activity.*

close by the Sun, traveling at fantastic speeds, do the ices become active and outgas large amounts of matter. As the comet journeys into the inner solar system, carbon monoxide ice vaporizes first and is replaced by jets of water vapor as the driving force behind the comet's growing brightness. Water vapor and gases stream outward, forming a tail millions of miles long that points away from the Sun due to the out-flowing solar wind.

The closest comet to approach the Earth was Lexell, which came within six times the distance to the Moon on July 1, 1770. The April 10, 837, encounter with Comet Halley was close enough for the Earth's gravity to disturb the comet's orbit. The first recorded impact of a comet occurred in the Tunguska Forest of northern Siberia on June 30, 1908. The tremendous explosion toppled and charred trees within a 20-mile radius. Tree trunks were splayed outward from the center of the blast like the spokes of a wheel. However, the explosion left no impact crater, suggesting a comet or stony asteroid airburst at an altitude of about 5 miles at a speed of 30,000 miles per hour.

The comet was relatively small, estimated between 100 and 300 feet wide. This explains why no astronomical sightings were made prior to the explosion. The estimated force of the blast was as powerful as a 15-megaton hydrogen bomb. Barometric disturbances were recorded over the entire world as the shock wave circled the Earth twice. The dust generated by the explosion produced unusual sunsets and bright skyglows observed at night over Europe within a few days of the event. The faint red glow was bright enough to read newspapers by it. If the impactor had exploded over a sizable city, the entire town and its suburbs would have been laid waste.

Almost all close encounters took astronomers completely by surprise, and not a single one was anticipated. To avoid the danger of an asteroid collision, the threatening body would first have to be tracked by telescopes and radars and its course plotted accurately so its orbit could be determined precisely. If an asteroid were found to be on a collision course with the Earth, astronomers could provide timely warnings to evacuate the threatened area. The rogue asteroid might also be nudged out of its earthbound trajectory by the detonation of nuclear warheads to one side. However, care must be taken not to shatter the asteroid, for what was once a single bullet would be turned into deadly buckshot and many fragments might cause more damage to the Earth.

STONES FROM THE SKY

Meteoroids are pieces broken off by constant collisions among asteroids. Due to the immense numbers of these fragments, meteorite falls are quite common. Daily, thousands of meteoroids enter the Earth's atmosphere. Occasional meteor showers can involve hundreds of thousands of stones, which almost invariably burn up during their passage through the atmosphere. The term *meteoroid* is generally used to refer to rocky matter in the solar system. Meteors, in contrast, are meteoroids that enter and burn up in the Earth's atmosphere. Meteorites, however, are meteors that actually survive to the Earth's surface.

Nearly 1 million tons of meteoritic material are produced annually. Much of it is suspended in the atmosphere, where it is responsible in part for making the sky blue. Fortunately, most meteors burn up upon entering the atmosphere. The remainder that fly through the atmosphere and rain down onto Earth can cause much havoc as meteorites crash into houses and automobiles.

Only meteors of a certain size are sufficiently large to travel all the way through the atmosphere without completely burning up. A meteoroid landing onto the Earth's surface is called a meteorite; the suffix *-ite* designates it as a rock. Meteorites composed of rock or iron do not appear to originate from the meteoroid streams created by the tails of comets but, instead, are fragments of asteroids chipped off by constant collisions. Some rare meteorites found in Antarctica are thought to be rocks blasted out of Mars by large impacts.

More than 500 meteorite falls occur each year. Most plunge into the oceans, which cover more than 70 percent of the Earth's surface, and accumulate on the seafloor. The majority of meteorites that fall onto land are slowed down by the breaking action of the atmosphere and only bury themselves a short distance into the ground. Not all meteorites are hot when they land because the lower atmosphere tends to cool them sufficiently, sometimes forming a thin layer of frost on their surfaces.

A spectacular meteorite fall of 3,000 stones at l'Aigle in the French province of Normandy in 1803 sparked the early investigation of meteorites. This spectacle was actually eclipsed 9 years earlier by a massive meteorite shower in Siena, Italy, on June 16, 1794. It was the most significant fall in recent times and spawned the modern science of meteoritics.

The earliest reports of meteorite falls were made by the ancient Chinese in the 17th century B.C. Chinese meteorites are rare, however. To date, no large impact craters have been recognized in China. One of the oldest meteorites that remains preserved in a museum is a 120-pound stone that landed near Alsace, France, on November 16, 1492. The largest known meteorite, called Hoba West, named for the farm it landed on, was found near Grootfontein, South Africa, in 1920 and weighed about 60 tons.

The largest meteorite find in the United States is the 16-ton Willamette Meteorite, which is as much as a million years old. It was discovered near Portland, Oregon, in 1902 and measured 10 feet long, 7 feet wide, and 4 feet high. One of the largest meteorites actually seen falling from the sky was an 880-pound stone that landed in a farmer's field near Paragould, Arkansas, on March 27, 1886. The heaviest observable stone meteorite landed in a cornfield in Norton County, Kansas, on March 18, 1948 and dug a pit 3 feet wide and 10 feet deep.

The iron meteorites are easily recognizable, although they represent only about 5 percent of all meteorite falls. They are composed of iron and nickel along with sulfur, carbon, and traces of other elements. Their composition is thought to be similar to that of the Earth's iron core. Indeed, they might have originated from the cores of large planetoids that disintegrated long ago. Due to their dense structure, iron meteorites have a good chance of surviving an impact, and most are found by farmers plowing their fields.

One of the best hunting grounds for meteorites happens to be on the glaciers of Antarctica, where the dark stones stand out in stark contrast to the white snow and ice. When meteorites fall onto the continent, they embed themselves in the moving ice sheets. At places where the glaciers move upward against mountain ranges, the ice evaporates, leaving meteorites exposed on the surface. Some meteorites that have landed on Antarctica are believed to have come from the Moon and from even as far away as Mars (Fig. 166) from large impacts that blasted out chunks of material and hurled them toward the Earth.

Perhaps the world's largest source of meteorites is the Nullarbor Plain, an area of limestone that stretches for 400 miles along the south coast of Western and South Australia. The pale, smooth desert plain provides a perfect backdrop for spotting meteorites, which are usually dark brown or black. Since very little erosion takes place, the meteorites are well preserved and are

Figure 166 *A meteorite discovered in Antarctica believed to be of Martian origin.*

(Courtesy NASA)

found just where they have landed. More than 1,000 fragments from 150 meteorites that fell during the last 20,000 years have been recovered. One large iron meteorite, called the Mundrabilla Meteorite, weighed more than 11 tons.

Stony meteorites are the most common type and make up more than 90 percent of all meteorite falls. However, because their composition is similar to the materials that make up the Earth and therefore erode easily, they are often difficult to locate. The meteorites are composed of tiny spheres of silicate minerals in a fine-grained matrix. The spheres are known as chondrules, from the Greek *chondros,* meaning "grain," and the meteorites themselves are therefore called chondrites. Most chondrites have a chemical composition believed to be similar to rocks in Earth's mantle, which suggests they were once part of a large planetoid that disintegrated eons ago. One of the most important and intriguing varieties of chondrites are the carbonaceous chondrites, which are among the most ancient bodies in the solar system. They also contain carbon compounds that might have provided the precursors of life on Earth.

The explosion of an asteroid in modern times was first observed by the pilot of a Japanese cargo plane over the Pacific Ocean about 400 miles east of Tokyo, Japan, on April 9, 1984. A ball-shaped cloud rapidly expanded in all directions, appearing much like a nuclear detonation, except the fireball or lightning that usually accompany nuclear explosions was not observed. Furthermore, an aircraft sent into the cloud to collect dust samples found no radioactivity. The mushroom cloud grew to 200 miles in diameter, rising from 14,000 to 60,000 feet in just two minutes. Apparently, the cloud formed from the explosion of an asteroid 80 feet in diameter, releasing the equivalent energy of a one-megaton hydrogen bomb.

If a meteor explodes as it streaks across the sky, it produces a bright fireball, called a bolide. The Great Fireball that flashed across New Mexico and nearby states on March 24, 1933, was as bright as the Sun. The meteoritic cloud grew to a towering plume in about five minutes. Some bolides are bright enough to be visible in broad daylight. Occasionally, their explosions can be heard on the ground and might sound like a thunderclap or the sonic boom of a jet aircraft. Every day, thousands of bolides occur around the world, but most go completely unnoticed.

IMPACT EFFECTS

When a large extraterrestrial body, such as an asteroid or comet nucleus, slams into the Earth, the impact can produce powerful blast waves, immense tsunamis, thick dust clouds, extremely toxic gases, and strong acid rains that can cause tremendous havoc. The tsunamis generated by a splashdown in the ocean are particularly hazardous to onshore and nearshore inhabitants. Perhaps the worst environmental hazards are produced by huge volumes of suspended sediment in the atmosphere from material blasted out of the crater along with vaporized asteroidal material. Furthermore, soot from continent-sized wildfires set ablaze by hot crater debris would clog the skies, causing darkness at noon.

This added burden would dramatically raise the density of the atmosphere and greatly increase its opacity, making conditions nearly impossible for sunlight to penetrate. Solar radiation would also heat the darkened, sediment-laden layers of the atmosphere and cause a thermal imbalance that could radically alter weather patterns, turning much of the land into a barren desert. Horrendous dust storms driven by maddening winds would rage across whole continents, further clogging the skies. So much damage would beset the Earth by major impacts that subsequent mass extinctions are considered a certainty.

A large asteroid impacting onto Earth could set the planet ringing like a bell. Large meteorite impacts can create so much disturbance in the Earth's crust that volcanoes and earthquakes could become active in zones of weakness. A massive impact in the Amirante Basin, 300 miles northeast of Madagascar, might have triggered India's great flood basalts, known as the Deccan Traps, when the subcontinent was headed toward southern Asia 65 million years ago. Quartz grains shocked by high pressures generated by a large meteorite impact found lying just beneath the immense lava flows might be linked to the impact.

Some geomagnetic reversals, whereby the Earth's magnetic poles switch polarities, appear to be associated with large impacts (Table 11). Magnetic reversals occurring 2.0, 1.9, and 0.7 million years ago coincided with unusual cold spells. Furthermore, the last two reversals correlated with the impacts of large meteorites on the Asian mainland and in the Ivory Coast region. Among the most striking examples of a large meteorite impact causing a magnetic reversal is the 15-mile-wide Ries Crater in southern Germany, which is about 14.8 million years old. The magnetization of the fallback material into the crater indicates that the geomagnetic field polarity reversed soon after the impact.

A single large meteorite impact or a massive meteorite shower would eject tremendous amounts of debris into the global atmosphere. This would block out the Sun for many months or years, possibly bringing down surface temperatures significantly to initiate glaciation. An asteroid apparently impacted onto the Pacific seafloor roughly 700 miles westward of the tip of South America about 2.3 million years ago. Geologic evidence suggests that the climate changed dramatically between 2.2 and 2.5 million years ago, when glaciers began to cover large parts of the Northern Hemisphere.

TABLE 11 COMPARISON OF MAGNETIC REVERSALS WITH OTHER PHENOMENA (DATES IN MILLIONS OF YEARS)

Magnetic Reversal	Unusual Cold	Meteorite Activity	Sea Level Drops	Mass Extinctions
0.7	0.7	0.7		
1.9	1.9	1.9		
2.0	2.0			
10				11
40			37–20	37
70			70–60	65
130			132–125	137
160			165–140	173

If a large meteorite entered the Earth's atmosphere, air friction would produce a brilliant meteor brighter than the Sun. The searing heat would scorch everything within miles around. Shock waves generated by the meteorite's blazing speed would be strong enough to knock people down more than 20 miles away. The impact would produce a rapidly expanding dust plume that would grow several thousand feet across at the base and extend several miles high. Most of the surrounding atmosphere would be blown away by the tremendous shock wave produced by the impact. The giant plume would turn into an enormous black dust cloud that punches through the atmosphere, like the mushroom cloud formed by the detonation of a hydrogen bomb (Fig. 167).

Heat produced by the compression of the atmosphere and impact friction that flings molten rock far and wide could set globalwide forest fires. The fires would consume some 80 percent of the terrestrial biomass, turning the planet into a smoldering cinder. The impact would also send aloft some 500 billion tons of sediment into the atmosphere. A heavy blanket of dust and soot would cover the entire globe and linger for months. A year of darkness would ensue under a thick brown smog of nitrogen oxide. Waters would be poisoned by trace elements leached from the soil and rock, and acid rains would be as corrosive as battery acid.

If the asteroid landed in the ocean, upon impact it would produce a conical-shaped curtain of water. Billions of tons of seawater would be splashed high into the atmosphere. The meteorite would instantly evaporate massive quantities of seawater, saturating the atmosphere with billowing clouds of

Figure 167 *Detonation of a hydrogen bomb, Pacific Proving Grounds.* (Courtesy U.S. Air Force)

steam. Thick cloud banks would shroud the planet, cutting off the Sun and turning day into night. The most massive tsunamis ever imagined would race outward from the impact site. The waves would completely traverse the world. When striking seashores, they would travel hundreds of miles inland, devastating everything in their paths, making such impacts among the most devastating, if rare, geologic hazards on Earth.

After discussing asteroid and comet impacts on the Earth and their effects on the planet and its life, the last chapter reveals the biologic consequences of those disasters as well as others such as climate cooling, volcanic eruptions, and magnetic reversals.

10

MASS EXTINCTIONS

THE LOSS OF LIFE

This chapter examines the extinction of species in Earth history, their causes and effects, and the perils species face in the world today. When considering all the great upheavals in the Earth throughout its long history, the fact that life has managed to survive to the present is truly remarkable. More than 99 percent of all species that have inhabited the planet since its creation have become extinct, so that those living today represent only a tiny fraction of the total.

As many as 4 billion species of plants and animals are thought to have existed in the geologic past. Most lived during the last 600 million years, a period of phenomenal evolution of species as well as tragic episodes of mass extinctions (Table 12). Therefore, the extinction of species has been almost as prevalent as their origination. The common denominator in all mass extinctions is that biologic systems were in extreme stress due to rapid and extreme changes in the environment.

HISTORIC EXTINCTIONS

Around 670 million years ago, thick ice sheets spread over much of the landmass during perhaps the greatest period of glaciation the Earth has ever

TABLE 12 RADIATION AND EXTINCTION OF SPECIES

Organism	Radiation	Extinction
Mammals	Paleocene	Pleistocene
Reptiles	Permian	Upper Cretaceous
Amphibians	Pennsylvanian	Permian-Triassic
Insects	Upper Paleozoic	
Land plants	Devonian	Permian
Fish	Devonian	Pennsylvanian
Crinoids	Ordovician	Upper Permian
Trilobites	Cambrian	Carboniferous & Permian
Ammonoids	Devonian	Upper Cretaceous
Nautiloids	Ordovician	Mississippian
Brachiopods	Ordovician	Devonian & Carboniferous
Graptolites	Ordovician	Silurian & Devonian
Foraminiferans	Silurian	Permian & Triassic
Marine invertebrates	Lower Paleozoic	Permian

known called the Varanger ice age. During this time, massive glaciers overran nearly half the continents for millions of years. All continents were assembled into a supercontinent, which might have wandered over one of the poles and collected a thick sheet of ice. That ice age dealt a deathly blow to life in the ocean. Many simple organisms vanished during the world's first mass extinction, when animal life was still scarce. The late Precambrian extinction decimated the ocean's population of acritarchs, a community of planktonic algae that were among the first organisms to develop elaborate cells with nuclei.

When glaciation ended, a rapid population growth ensued, with a diversification of species that has not been equaled since. An explosion of species that represented nearly every major group of marine organisms set the stage for the evolutionary development of more modern life-forms. By the time the Precambrian era came to a close, the seas contained large populations of widespread and diverse species. Never before or since had such a diversity of species existed, and some of the strangest animals ever known populated the planet (Fig. 168).

The fossil record from that era is dominated by numerous unusual creatures, many of which probably arose by adapting to highly unstable conditions. As a consequence of overspecialization, a major extinction of species

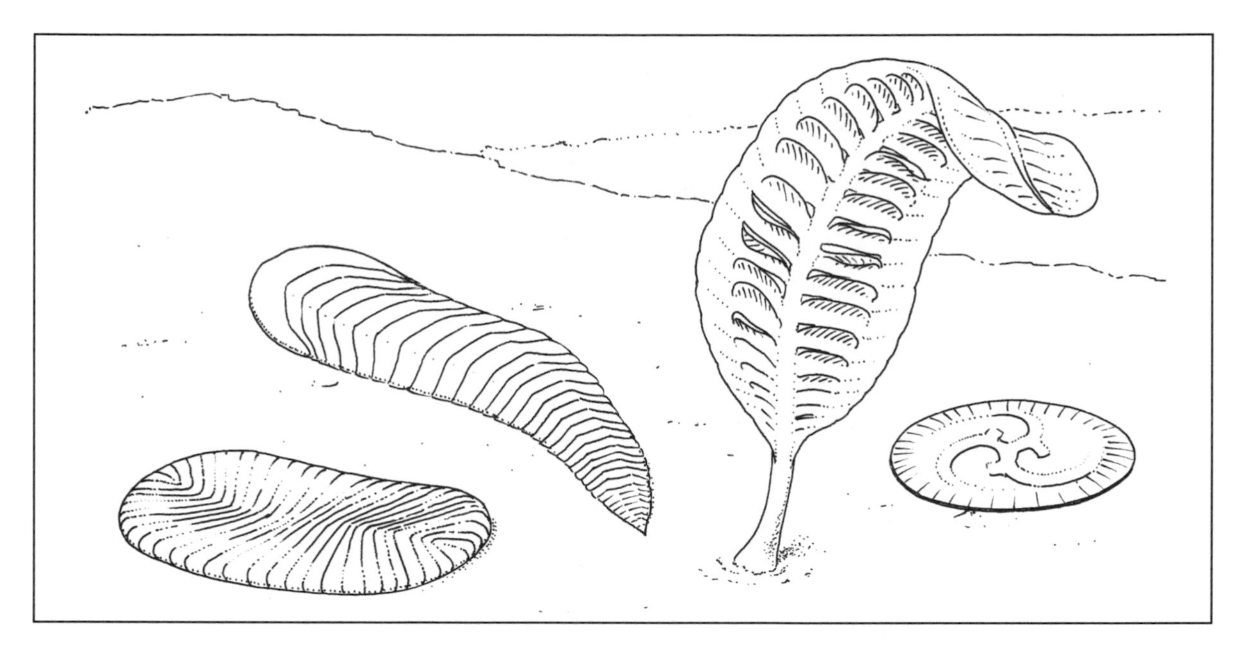

Figure 168 *The late Precambrian marine fauna.*

took place at the end of the Precambrian 570 million years ago. Species that survived the extinction were quite different from those that did not. These new life-forms flourished in the warm climate and participated in the greatest explosion of species in geologic history. Most of these novel species were not related to modern forms.

The Phanerozoic eon, from the Cambrian period beginning 570 million years ago to the present, witnessed several mass extinctions that eliminated vast arrays of species. The Cambrian was an evolutionary heyday. The first complex animals with exoskeletons exploded onto the scene, filling the seas with a rich assortment of life. During the great Cambrian explosion, species diversity was at an all-time high. The continents, which before were in the polar regions, drifted into warmer areas around the equator. The increased temperatures and warm Cambrian seas spurred a rapid evolution of species. Many experimental organisms came into existence at this time, none of which have any modern counterparts.

No sooner had the Cambrian started when a wave of extinctions decimated a huge variety of newly evolved species. The extinctions, which were among the most severe in Earth history, eliminated more than 80 percent of the marine animal genera. The die-offs wiped out most major groups, paving the way for the ascendancy of a famous class of invertebrates called the trilobites (Fig. 169). These were primitive crustaceans and ancestors of today's horseshoe crab. They were among the first animals well protected

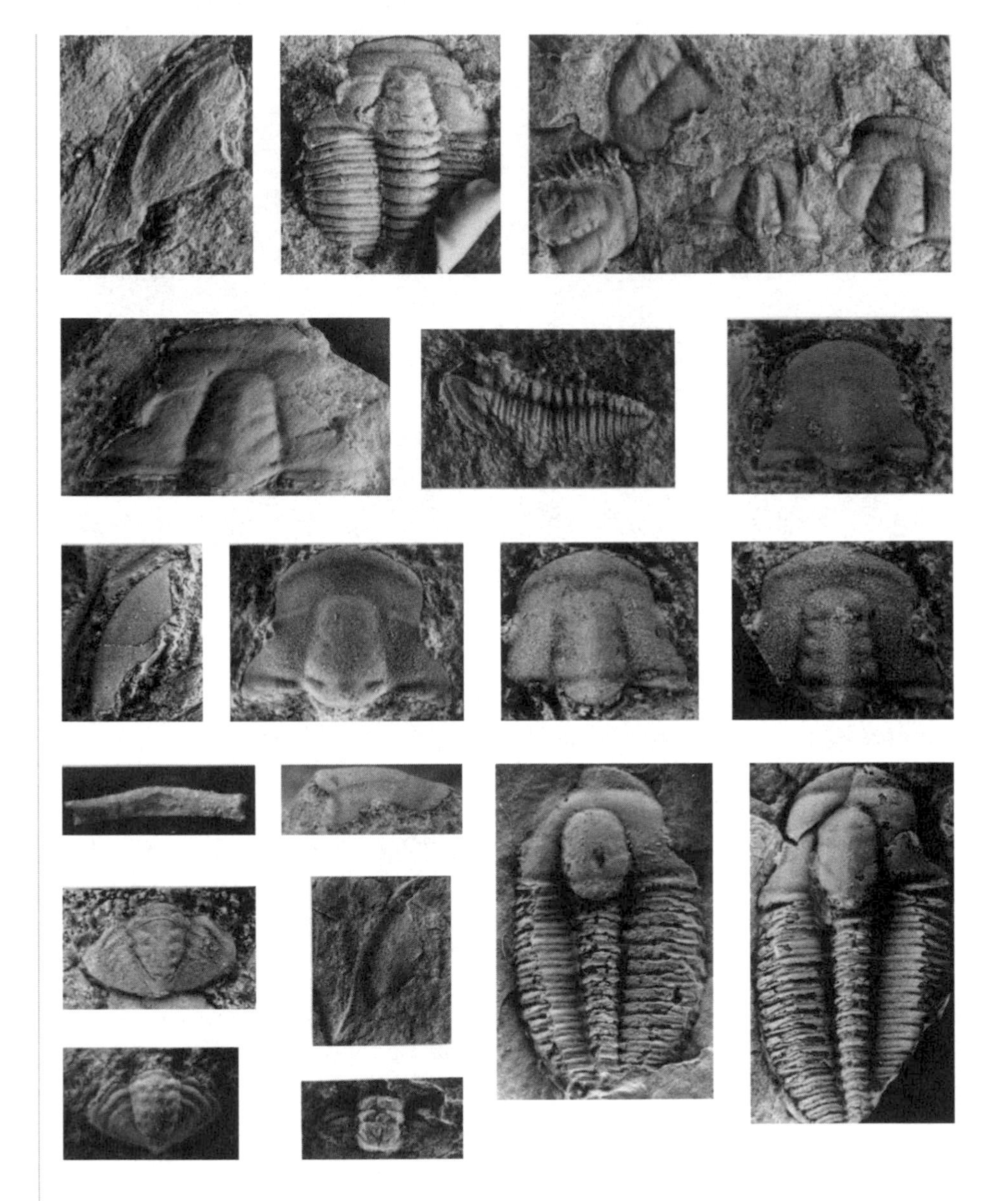

Figure 169 *Trilobite fossils of the Carrara Formation in the Southern Great Basin, California-Nevada.*

(Photo by A. R. Palmer, courtesy USGS)

with hard shells and became the dominant species for the next 100 million years.

A second mass extinction eliminated some 100 families of marine animals at the end of the Ordovician period, about 440 million years ago. Many animal groups exploded during the Ordovician. The number of marine families bloomed from 160 to 530, while the number of genera ballooned to nearly 1,600. Several mountain ranges appeared at the same time, suggesting a link between geologic and biologic revolutions.

At the end of the Ordovician, glaciation reached its peak, with ice sheets radiating outward from a center in North Africa. The frigid conditions resulted in a mass extinction that eliminated some 100 families of marine animals. Tropical species were the hardest hit because the tropics are more sensitive to fluctuations in the environment. Among those that went extinct were many species of trilobites. Before the extinction, they accounted for about two-thirds of all species. However, they represented only one-third thereafter. The graptolites, which were a strange animal species resembling a conglomeration of floating stems and leaves, also became extinct.

Another major extinction took place during the middle Devonian, about 365 million years ago, when many tropical marine groups simultaneously disappeared, possibly due to a period of climatic cooling. The extinction appears to have extended over a period of 7 million years. Corals (Fig. 170) which were prolific limestone reef builders, suffered an extinction from which they never fully recovered. The extinction did not identically affect all species that shared the same environments, however. Many fauna survived the onslaught due to the scarcity of reef builders and other warm-water species prone to extinction. When the corals disappeared due to the recession of the sea in which they once thrived, they were replaced by sponges and algae in the late Paleozoic. Ninety percent of the brachiopod families, which had two clamlike shells fitted face to face that opened and closed with simple muscles, also died out at the end of the Devonian.

No major die-off occurred during the widespread Carboniferous glaciation, which enveloped the southern continents around 330 million years ago. The relatively low extinction rates were credited to a limited number of extinction-prone species following the late Devonian extinction. When the glaciers departed, the first reptiles emerged to displace the amphibians as the dominant land vertebrates. The climate of the Tropics became more arid, and the swamplands started to disappear. Land once covered with great coal swamps began to dry out as the climate grew colder. The climate change set off a wave of extinctions that wiped out virtually all the lycopods including club mosses and scale trees, which dominated the world's forests at that time (Fig. 171).

The greatest mass extinction event occurred at the end of the Permian period, 250 million years ago. The extinction probably spanned no more than a million years. It coincides with a massive eruptive outpouring of basalts in Siberia, forming huge stairlike flows called traps. Half the families of marine organisms, comprising more than 95 percent of all known species, abruptly disappeared. About 85 percent of the species living in the ocean died out. On land, more than 70 percent of the vertebrates went extinct. One-third of insect orders ceased to exist, marking the only mass extinction insects have

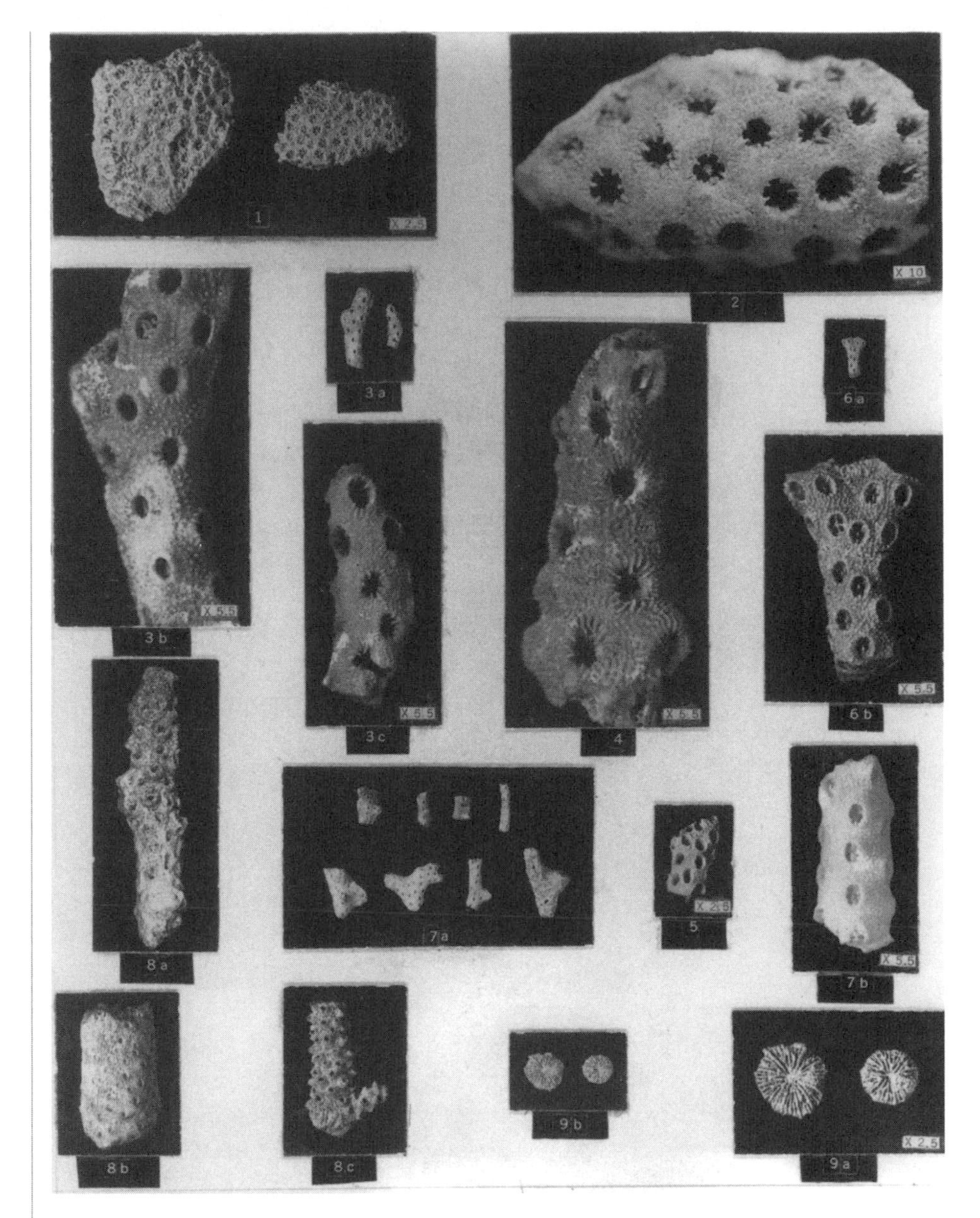

Figure 170 *Fossil corals from Bikini Atoll, Marshall Islands.*

(Photo by J. W. Wells, courtesy USGS)

ever undergone. So many plants succumbed to extinction that fungi briefly ruled the continents.

The extinction followed on the heels of a late Permian glaciation. Marine invertebrates that managed to escape extinction were forced to live in a narrow margin near the equator. Corals, which require warm, shallow waters, were particularly hard hit, as evidenced by the lack of coral reefs in the early part of the Triassic period. Brachiopods and crinoids, early relatives of starfish that grew on long stalks and had their golden age in the Paleozoic, were relegated to minor roles during the following Mesozoic era. The trilobites, which were extremely successful during the Paleozoic, completely died out at the end of the era.

At the end of the Triassic, about 210 million years ago, 20 percent or more families, mostly terrestrial animals, began dying off in record numbers, eliminating some 50 percent of all species. The extinction followed the initial breakup of Pangaea, when massive floods of basalt lava covered the continents, perhaps the largest outpouring known in Earth history. The extinction occurred over a period of less than a million years and was responsible for killing off nearly half the reptile families. Marine fauna, including foraminifers, ammonoids, bivalves, bryozoans, echinoids, and crinoids, experienced global-scale extinctions. The Triassic crisis also eliminated tropical reef corals, possi-

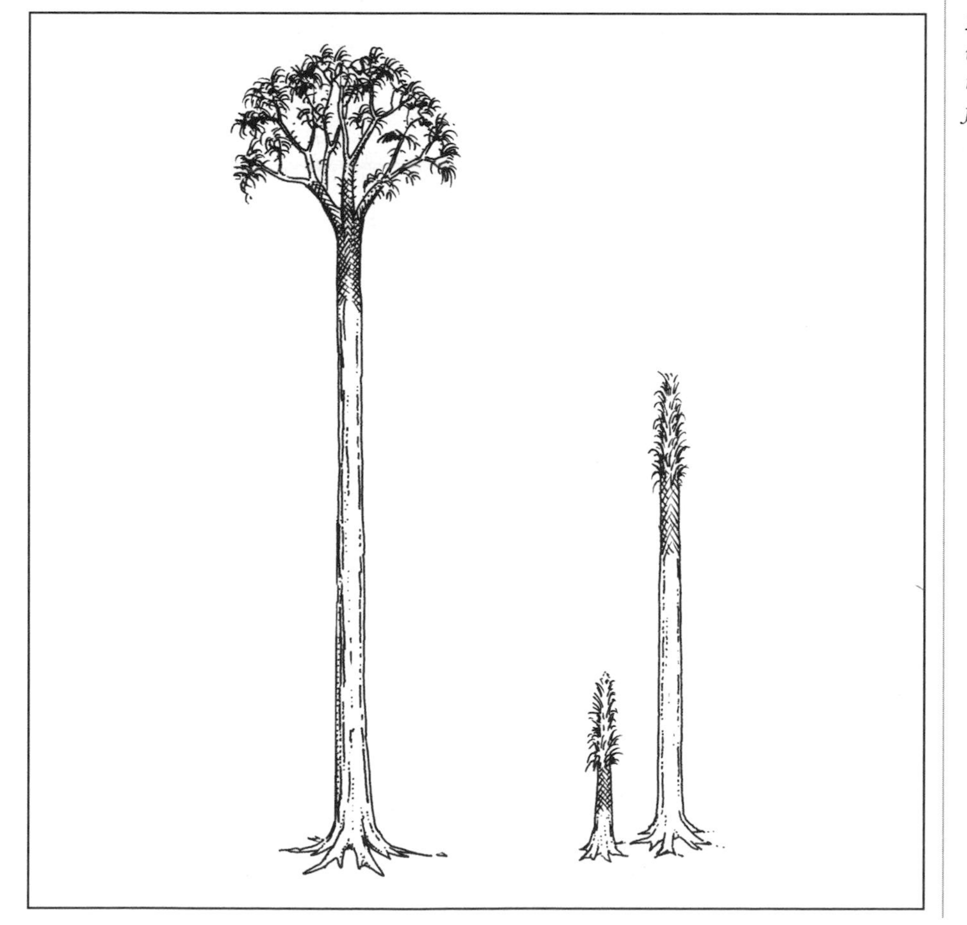

Figure 171 *The lycopods were one of the dominant trees of the Paleozoic forests.*

Figure 172 *Allosaurs were among many large dinosaur species to become extinct at the end of the Jurassic.*

(Courtesy National Museums of Canada)

bly due to a colder climate. The extinctions forever changed the character of life on Earth and paved the way for the rise of the dinosaurs.

Many large dinosaur families, including brontosaurs, stegosaurs, and allosaurs (Fig. 172), were cut down at the end of the Jurassic period 135 million years ago. Following the extinction, the population of small animals exploded as species occupied niches vacated by the large dinosaurs. Most of the surviving species were aquatic creatures confined to freshwater lakes and marshes along with small, land-dwelling animals. Many of the small, nondinosaur species were the same ones that survived the next great mass extinction, probably because of their large populations and their ability to find places to hide.

The most popularized extinction took place at the end of the Cretaceous period, 65 million years ago. The famous dinosaurs along with 70 percent of all other known species, mostly marine animals, suddenly vanished. Archaic mammals and birds, the dinosaurs' closest living relatives, suffered the same fate. The success of the dinosaurs is exemplified by their extensive range. They occupied a wide variety of habitats and dominated all other terrestrial animals. About 500 dinosaur genera thus far have been classified. Because the dinosaurs were not the only species to become extinct, something in the environment must have made survival impossible. Yet most mammals were not seriously affected.

The ammonites, cephalopods with a large variety of coiled shells up to 7 feet across (Fig. 173), were fantastically successful in the warm Cretaceous

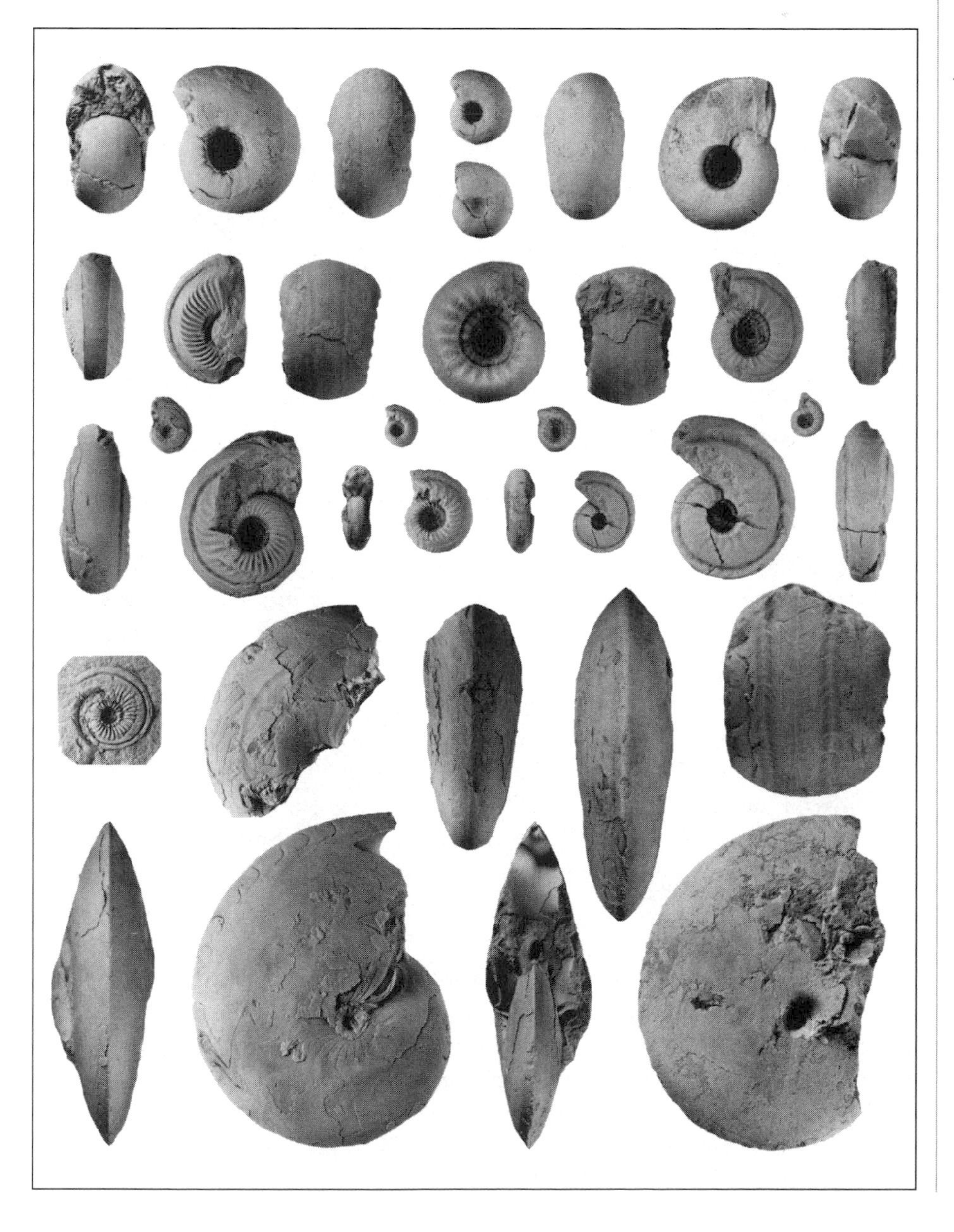

Figure 173 *A variety fossil ammonite shells.*

(Photo by M. Gordon Jr., courtesy USGS)

seas. After surviving the critical transition from Permian to Triassic and recovering from serious setbacks in the Mesozoic, the ammonites suffered final extinction at the end of the Cretaceous when the recession of the seas reduced their shallow-water habitats worldwide. Other major marine groups that disappeared included the rudists, which were large, coral-shaped clams. Half the bivalve genera, including clams and oysters, also died out. All marine reptiles, except the smallest of turtles, became extinct at the end of the period. Ninety percent of all plankton species living in the surface waters of the ocean also died out.

Following the great Cretaceous extinction, life experienced an evolutionary lag lasting up to several million years. Afterward, the mammals began to diversify rapidly, sometimes giving rise to unusual forms. Extremes in climate and topography during the Cenozoic, the era that came after the age of the dinosaurs, produced a greater variety of living conditions than any other equivalent span of Earth history. The rigorous environments presented many challenging opportunities. The extent to which plants and animals invaded diverse habitats was truly remarkable. The invasion of new habitats during repopulation following a major extinction is the major source of evolutionary opportunities. As a response to those opportunities, species show a burst of evolutionary development that gives rise to a great diversity.

A sharp extinction event occurred at the end of the Eocene epoch, 37 million years ago, when the Earth plunged into a colder climate. The extinction wiped out most archaic mammals, which were large, grotesque-looking animals. A significant fall in sea level resulting from a major expansion of the Antarctic ice sheet led to another extinction about 11 million years ago. These cooling events removed the most vulnerable species, leaving those living today highly robust, having withstood the extreme environmental swings over the last 3 million years when glaciers spanned much of the Northern Hemisphere.

The latest mass extinction was one of the strangest of all. It occurred toward the end of the last Ice Age, when giant mammals called megaherbivores disappeared, including the saber-toothed cats, ground sloths, mastodons, and woolly mammoths (Fig. 174). North America lost about three-quarters of all animal genera weighing more than 100 pounds. Australia suffered the most severely of all continents, losing every terrestrial vertebrate species larger than a human as well as many smaller mammals, reptiles, and flightless birds.

Why these large animals disappeared at this time after surviving several episodes of glaciation over the previous 3 million years remains a mystery. When the glaciers began to retreat around 16,000 years ago, a major readjustment in the global environment might have disrupted food supplies, causing several megaherbivore species to become extinct. Moreover, humans were becoming efficient hunters of large game and might have slaughtered some giant mammals to extinction.

Figure 174 Mammoth and many other large mammals became extinct at the end of the last Ice Age, about 11,000 years ago, at the same time humans migrated into their habitats.

CAUSES OF EXTINCTIONS

Most extinctions of the past appeared to have coincided with major planetary changes brought on by tectonic forces and lowering sea levels. Global cooling caused many extinctions because climate is possibly the most important factor influencing species diversity. As the world's oceans cool, mobile species tend to migrate to the warmer regions of the Tropics. Species attached to the ocean floor and unable to move or those trapped in enclosed basins are generally the hardest hit by extinction. Only species previously adapted to cold conditions still thrive in today's oceans. Most are plant eaters that tend to be generalized feeders consuming a variety of foods.

Much mass extinction coincided with periods of glaciation because temperature is perhaps the single most important factor limiting the geographic distribution of species. Certain species, such as corals, can survive only within a narrow range of temperatures. During warm interglacial periods, species invade all latitudes. When glaciers advance across continents and oceans, temperatures drop and species are forced into warmer regions, where limited habitats exist. The intense competition for habitat and food severely limits species diversity and therefore the total number of species.

Not all climate cooling resulted in glaciation, however. Nor did all extinctions follow a drop in sea level caused by growing glaciers. During the Oligocene epoch, which began about 37 million years ago, seas that overrode the continents drained away as the ocean withdrew to one of its lowest levels in several hundred million years. Although the sea level remained depressed for

5 million years, little or no excess extinction of marine life occurred. Therefore, crowding conditions brought on by lowering sea levels cannot be responsible for all extinctions. Furthermore, during many mass extinctions, the sea level was not much lower than it is today.

In the final stages of the Cretaceous period, when the dinosaurs walked the Earth, the level of the ocean began to drop, seas departed from the land, and the temperatures in a broad tropical ocean belt called the Tethys Sea began to fall. The change in sea level might explain why the Tethyan species that were the most temperature sensitive endured the heaviest extinction at the end of the period. Species that were amazingly successful in the warm waters of the Tethys suffered total decimation when temperatures dropped. After the extinction, marine species acquired a more modern appearance as ocean bottom temperatures continued to plummet.

During the past 250 million years, 11 distinct episodes of flood basalt volcanism occurred throughout the world (Fig. 175 and Table 13). Catastrophic volcanic episodes apparently took place at intervals of 200 million years, with lesser events spaced roughly 30 million years apart. This corresponded to an apparent 26- to 32-million-year periodicity of extinction. These were relatively short-lived events, with major phases spanning less than 3 million years. The timing of these volcanic outbreaks correlates well with the occurrence of mass extinctions of marine species. Extensive volcanism occurring at the end of the Cretaceous might have killed off the dinosaurs and most other land species along with many marine species.

A large number of volcanoes erupting over a long interval could lower global temperatures by injecting massive quantities of volcanic ash and dust

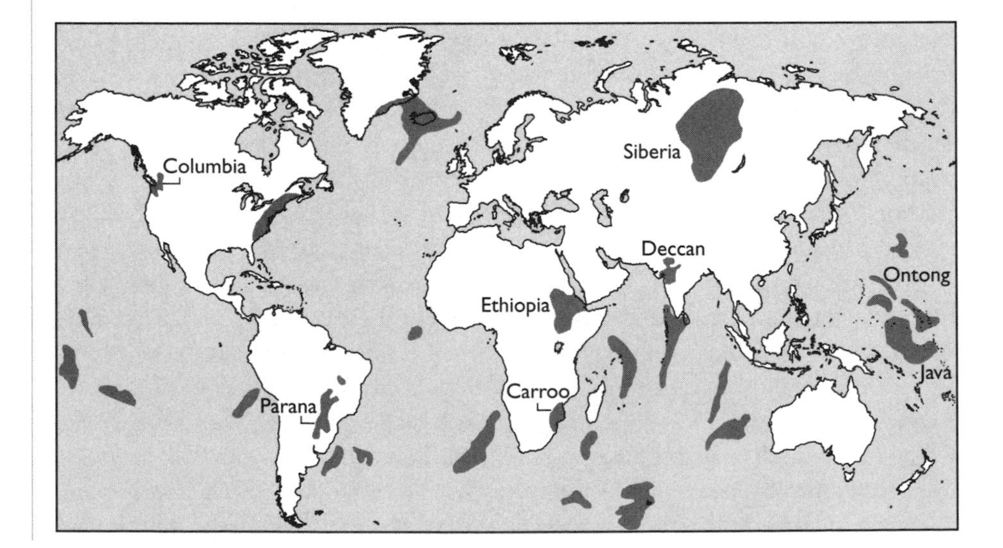

Figure 175 *Areas affected by flood basalt volcanism.*

TABLE 13 FLOOD BASALT VOLCANISM AND MASS EXTINCTIONS

Volcanic Episode	Million Years Ago	Extinction Event	Million Years Ago
Columbian River, USA	17	Low-mid Miocene	14
Ethiopian	35	Upper Eocene	36
Deccan, India	65	Maastrichtian	65
		Cenomanian	91
Rajmahal, India	110	Aptian	110
Southwest African	135	Tithonian	137
Antarctica	170	Bajocian	173
South African	190	Pliensbachian	191
E. North American	200	Rhaectian/Norian	211
Siberian	250	Guadalupian	249

into the atmosphere. Heavy clouds of volcanic dust have a high albedo and reflect solar radiation back into space, thereby shading the Earth. The lowered global temperatures could cause mass extinctions of plants and animals by reducing the rate of global photosynthesis. However, plants can survive short-term catastrophes far better than animals by wilting during droughts, resprouting from roots, and maintaining prolonged dormancy as seeds.

Acid rain from extensive volcanic activity a hundred times more intense than at present could cause widespread destruction of terrestrial and marine species by defoliating plants and raising the acidity level of the ocean. Acid gases spewed into the atmosphere might deplete the ozone layer, allowing deadly ultraviolet radiation from the Sun to bathe the planet. Volcanic eruptions also affect the climate by altering the composition of the atmosphere. Large volcanic eruptions spew so much ash and aerosols into the atmosphere they not only block sunlight but also absorb solar radiation, which heats the atmosphere, causing thermal imbalances and unstable climatic conditions.

Geologic evidence taken from sequences of volcanic rock on the ocean floor, which record the polarity of the Earth's magnetic field when the rocks cool and solidify, shows that the geomagnetic field has reversed often in the past. After a long, stable period of hundreds of thousands of years, the magnetic field strength gradually decays over a short period of several thousand years. At some point, it collapses altogether. A short time later it regenerates, half the time with opposite polarity.

A comparison between magnetic reversals with variations in the climate often shows a striking agreement. Furthermore, certain magnetic reversals coincide with the extinction of species. Magnetic field reversals might also have been responsible for periods of glaciation. Reversals in the geomagnetic field and excursions of the magnetic poles appear to correlate with periods of rapid cooling and extinction of species. For example, the Gothenburg geomagnetic excursion occurred about 13,500 years ago in the midst of a longer period of rapid global warming toward the end of the last Ice Age. It resulted in plummeting temperatures and advancing glaciers for a thousand years, apparently caused by a weakened magnetic field.

Ten or more major meteorites have impacted over the last 600 million years, some of which coincide with mass extinctions. When a large asteroid or comet slams into the Earth, a huge explosion lifts massive amounts of sediment into the atmosphere that shuts out the Sun. Darkness might last several months, halting photosynthesis and eliminating near-surface phytoplankton in the ocean. The effects of these extinctions would cascade up the food chain, killing off large marine and terrestrial species as well. A massive bombardment of meteoroids or comets might strip away the upper atmospheric ozone layer, leaving all species on the surface vulnerable to the Sun's deadly ultraviolet rays. The exposure would kill terrestrial plants and animals along with primary producers in the surface waters of the ocean.

A controversial theory to explain the extinction of the dinosaurs contends that a 6- to 9-mile-wide asteroid struck the Earth and gouged out a crater 100 miles in diameter. Boundary rocks between the Cretaceous and Tertiary periods throughout the world (Fig. 176) contain a thin layer of fallout material composed of mud with shock-impact sediments, microspherules produced by impact melt, organic carbon from massive forest fires, the mineral stishovite found only at known impact sites, meteoritic amino acids, and an unusually high iridium content. The geologic record holds clues to other large meteorite impacts associated with iridium anomalies that coincide with extinction episodes. Therefore, giant impacts might have had a hand in the initiation and extermination of species throughout geologic history.

EFFECTS OF EXTINCTIONS

Since life first appeared on Earth, gradual die-offs of species, called background or normal extinctions, have always been a common occurrence. Major extinction events are punctuated by periods of lower extinction rates, and species have regularly come and gone even during optimum conditions. However, mass extinctions are not simply intensifications of processes operating during

Figure 176 *Geologists point out the Cretaceous-Tertiary boundary at Browie Butte Outcrop, Garfield County, Montana.*

(Photo by B. F. Bohor, courtesy USGS)

these background periods. Survival traits developed during times of lower extinction rates become irrelevant during mass extinctions. This suggests that mass extinctions might be less discriminating with respect to the environment than normal extinctions. Therefore, different processes might be operating during times of mass extinction than those operating during normal extinctions. Furthermore, the same types of species that succumb to mass extinctions also succumb to background extinctions—only a lot more of them.

Species that survive mass extinction are particularly hardy and resistant to subsequent random changes in the environment. They tend to occupy large geographic ranges that contain many groups of related species. However, just because a species survives extinction does not always mean it was superior or better suited to its environment than species that died out. Species that became extinct might have been developing certain unfavorable traits during background times. This could occur even within generations of the same organisms, as daughter species develop better survival skills and replace their parent species. In this case, those characteristics that permit a species to live successfully during normal periods become irrelevant when major extinction events occur.

After mass extinction, species that survive radiate outward to fill vacated habitats, spawning the development of entirely new species. These new species

Figure 177 Marine fauna of the early Cambrian.

might develop novel adaptations that give them a survival advantage over other species. These adaptations might lead to exotic organisms that prosper during intervals of normal background extinctions but, because of overspecialization, are incapable of surviving mass extinction.

The geologic record seems to imply that nature is constantly experimenting with new forms of life (Fig. 177). When one species fails, it becomes extinct, never to appear again. Once a species becomes extinct, it is lost forever. Odds against its unique combination of genes reappearing are astronomical. Thus, evolution appears to run in one direction. Although it perfects species to live at their optimum in their respective environments, it can never go back to the past. However, convergent evolution provides the possibility for a species to resemble an entirely different species physically only because they share the same niche.

Extinction reduces the number of different species as well as the total number of species. Afterward, the biologic system seems to be temporarily immune to random cataclysms. Species that survive mass extinction are particularly hardy and resilient toward subsequent environmental changes. Furthermore, after a major extinction event, few species are left to die out. Therefore, until many species have evolved, including extinction-prone types, any intervening catastrophes would have comparatively little effect.

Following each extinction, the biologic world requires a recovery period before it is again ready to face another major extinction event. Each time a mass extinction occurs, it resets the evolutionary clock, as though life were forced to start anew. After the great extinction that ended the Paleozoic 250 million years ago, which left the world almost as devoid of species as when the era began, life witnessed many remarkable advancements. Species that survived the extinction were similar to populations living today. Many of these same species survived the end-Cretaceous extinction, suggesting they might have perfected survival characteristics that other species lacked.

As the world was recovering from the extinction at the end of the Paleozoic, many regions of the ocean became filled with numerous specialized organisms. The overall diversity of species rose to unprecedented heights. However, instead of evolving entirely novel forms, such as those that developed during the Cambrian explosion, species that survived the end-Paleozoic extinction developed morphologies or body styles based on simple skeletal types, producing few experimental life-forms.

MODERN EXTINCTIONS

Humans came into existence during the greatest biologic diversity in the history of the planet, when 70 percent of today's species evolved. We are a relatively new species on a geologic time scale (Fig. 178), especially when compared with others, some of which have been in existence for hundreds of millions of years. Within the past few thousand years, people have radiated into all lands. Within the last few hundred years, human populations have swelled a thousandfold. We are the most adaptable species, capable of living in diverse environments, often nudging out other species in competition.

Humans are the only creatures on Earth that have forced the extinctions of large numbers of other species. We have been called the "human volcano" because our influences on the environment are global just like major volcanic eruptions. Such convulsions have been cited as causes of mass extinction of species. Major changes in the Earth's critical cycles brought on by human interference could spell a catastrophe for all humankind as well as for the rest of the living world.

For the first time in geologic history, plants are being extinguished in tragic numbers. If current trends continue, a significant number of plant species are likely to become extinct. In the United States alone, as much as 10 percent of the nation's plant species are destined for extinction. Plants are at risk of extinction from forest destruction, expansion of agriculture, and the spread of urbanization. More than 1,000 domestic species of plants and animals are either endangered or threatened with extinction. Plants and animals not directly beneficial to humans are crowded out, as growing populations

Figure 178 *Geologic time spiral depicting the geologic history of the Earth.*

(Courtesy USGS)

continue to squander the Earth's space and resources and contaminate the environment with pollution (Fig. 179).

Everywhere humans have gone, they have wiped out entire species of mammals, birds, reptiles, fish, and other life-forms, including indigenous peoples. Between 1600 and 1900, during a period of extensive maritime exploration, humans eliminated 75 known species, mostly birds and mammals. Passenger pigeons that once darkened the skies over North America by the billions became totally extinct by 1914 due to overhunting and habitat destruction. We continue to destroy life on every continent and island we inhabit, except on a much larger scale than our forebears, simply because we have filled the world with so many more of us.

Wetlands are among the richest plant and animal communities in the world, supporting large numbers of diverse species. Many wetlands in the United States, such as the great Florida Everglades, have been rapidly modified by human activities. Diking and filling of wetlands have eliminated habitats of fish and waterfowl. The introduction of exotic species has transformed the composition of aquatic communities. The reduction of freshwater inflow has changed the dynamics of plant and animal communities in the wetlands. In addition, urban and industrial wastes have contaminated sediments as well as organisms. The continued disposal of toxic wastes and the further reduction of freshwater inflows greatly alter wetland water quality and biologic communities.

The world's wetlands are drained to provide additional farmland. Nearly 90 percent of recent wetland losses in the United States have been for agricultural purposes. Woodland marshes are disappearing at an alarming rate of more than 1,000 acres a day (Fig. 180). As sea levels continue to rise due to higher global temperatures, 80 percent of the U.S. coastal wetlands and estuaries could be lost by the middle of this century.

The urgent need to feed growing populations is the major reason the developing countries drain wetlands. Short-term food production has obscured the long-term economic and ecologic benefits of preserving wetland habitats. The disappearance of the wetlands is responsible for the loss of local fisheries and breeding grounds for marine species and wildlife. In many cases, wetland destruction is irreversible.

Fish species are rapidly disappearing throughout the world due to deforestation, which causes increased sedimentation, and acid rain, which acidifies lakes and streams. The oceans are not immune, either. Once large fisheries have

Figure 179 *Exxon workers using high-pressure cold water to rinse crude oil from a beach in Prince William Sound, Alaska.*

(Photo by J. Bauermeister, courtesy USGS)

Figure 180 *Swan Lake, a state refuge for waterfowl in Walworth County, South Dakota, dried out in 1974 due to drought conditions. Severe wind erosion occurred on about half the lake's 2,300-acre dry bed.*

(Photo by P. Kuck, courtesy USDA Soil Conservation Service)

been seriously depleted to feed growing human populations. Even the great sharks, which have been extremely successful predators for the last 400 million years, are succumbing to human overfishing.

Acid rain is a growing threat to the environment. It is especially harmful to aquatic organisms, as most species cannot tolerate high acidity levels in their environments. In seawater, the damage is due to nitrogen oxides in acid rain. Nitrogen is a nutrient that promotes the growth of floating algae (Fig. 181), which blocks sunlight and halts photosynthesis below the water's surface. When the algae die and decompose, bacteria deplete the water of its dissolved oxygen, which in turn suffocates other aquatic plants and animals.

The ocean is also contaminated from widespread increases in nitrate levels along with higher concentrations of toxic metals, including arsenic, cadmium, and selenium. The main factors contributing to these increases are fertilizer, herbicide, and pesticide runoff along with acid rain, which dissolves heavy metals in the soil. Moreover, riverine fisheries have been damaged by increased sedimentation from erosion and deforestation in catchment areas that supply water for rivers and aquifers.

In many areas, such as Bermuda, the Virgin Islands, and Hawaii, urban development and sewage outflows have led to extensive overgrowth by thick mats of algae that suffocate and eventually kill coral by supporting the growth of oxygen-consuming bacteria. The coral are particularly at risk during the

Figure 181 *Rafts of blue-green algae lying in a pool in Indian Canyon, Duchesne County, Utah.*

(Photo by W. H. Bradley, courtesy USGS)

Figure 182 *Tanapag Lagoon, Saipan, Mariana Islands, showing transition from barrier reef in the foreground to fringing reef at the left background.*

(Photo by P. E. Cloud Jr., courtesy USGS)

winter, when the algal cover on shallow reefs is extensive. This results in the loss of living coral and the eventual destruction of the reef by erosion.

Increasing ocean temperatures cause bleaching of many reefs, turning corals deathly white due to the expulsion of symbiotic (helpful) algae from their tissues. The algae aid in nourishing the corals, and their loss poses a great danger to the reefs. Foraminifera, which are marine plankton and important players in the global carbon cycle and food chains, are suffering a similar bleaching effect.

Tropical coral reefs (Fig. 182) are also centers of high biologic productivity. Their fisheries provide a major food source for the tropical regions. Unfortunately, the spread of tourist resorts along coral coasts in many parts of the world harms the productivity of these areas. Such developments are almost always accompanied by increased sewage dumping, overfishing, and physical damage to the reef by construction, dredging, dumping, and landfilling. Reefs are also destroyed to provide tourists with curios and souvenirs. In addition, upland deforestation chokes off coral reefs with eroded sediments transported to the sea by rivers.

Birds are also at risk. Humans have caused a significant number of all bird species to become extinct over the last few centuries due to overhunting, introduction of competing species, and destruction of habitat. Half the Hawaiian bird population has collapsed from overhunting and destruction of native forests following Polynesian habitation some 1,600 years ago. In more recent times, the bird population has been further reduced by 15 percent. This situation is typical of the impact of human settlement in island communities, which are especially vulnerable.

An estimated 2,000 bird species, or about one-fifth of all bird species that existed a few thousand years ago, have fallen victim to prehistoric exterminations. Today, as much as 20 percent of the bird species are endangered or at imminent risk of extinction. Large, flightless birds, such as the dodo and great auk, which were driven into extinction by people, are particularly at risk from human interference. Island birds are frequently flightless because they no longer need to take to the air to escape predators. Species inhabiting islands are extremely vulnerable to humans simply because they have nowhere to flee. Animals living on islands often develop unique characteristics apart from their mainland relatives that make them particularly at risk to human activities.

Tropical rain forests throughout the world are also disappearing at alarming rates. Tropical forests in the Western Hemisphere have shrunk about one-third, and those of Africa have been reduced by as much as 75 percent since 1960. These regions are the wintering grounds for migratory birds from the Northern Hemisphere. The disappearance of the rain forests could mean a decline in bird populations for the northern countries besides countless other species that inhabit the forests themselves. Already, birds are starting to disappear in alarming numbers. In the same way that canaries are used to detect poisonous gas in coal mines, bird deaths could be offering us a warning that our planet is in peril.

Birds are not the only species in trouble. The African elephant and black rhinoceros along with other large mammals are in danger of extinction due to a greedy ivory trade and human encroachment into their habitats. These large herbivores actually improve their environment by opening forests for grass undergrowth, which increases productivity and accelerates nutrient recycling. The cleaner forests are also much less vulnerable to forest fires. Unfortunately, with the elimination of these animals, their favorable environmental impacts are reversed, restricting the habitats of smaller herbivores, which follow their larger cohabitants into extinction.

THE WORLD AFTER

The extinctions of the past were caused by natural phenomena, including changing climatic and environmental conditions. However, present-day extinctions are caused by destructive human activities. Throughout the world, the die-off of species is thousands of times greater than the natural background extinction rate prior to the appearance of humans. We are growing so explosively and destroying the environment so extensively that species of plants and animals are perishing in tragically large numbers. As human populations continue to expand out of control and alter the natural environment, they have the potential of reducing biologic diversity to its lowest level since the great extinction of the dinosaurs and three-quarters of all known species 65 million years ago.

The number of species living today ranges from 5 million to as many as 30 million. Only about 1.4 million species have been formally classified, however. Therefore, more than 90 percent of all species have yet to be described and thus are unknown to science. Biologists are in a desperate race to classify as many species as possible before many more are gone. Most species go about their lives completely unnoticed. Many play critical roles in food chains and make important nutrients available to higher organisms. Simple creatures, including bacteria, fungi, and plankton, which make survival possible for all other species, make up more than 80 percent of the Earth's biomass (the total weight of living matter). Moreover, marine phytoplankton (Fig. 183) produce most of the breathable oxygen available on the planet. These are the organisms that hold nature in balance. If they go, so do we.

The interrelationships among species and between them and their environments is highly complex and difficult to comprehend. However, what is becoming more apparent is that the destruction of large numbers of the world's species would not only lower species diversity but allow species commonly called pests to flourish because their natural predators would be eliminated. By destroying beneficial predators, we allow pests and parasites to flourish, thereby upsetting the balance of nature. Therefore, the destruction of large numbers of species would leave us with a world entirely different from the one we inherited. Indeed, scientists doubt that human beings could survive in such a world because biologic support systems such as critical food chains would be severely damaged.

Figure 183
Phytoplankton such as coccolithphore help maintain living conditions on Earth.

The current extinction rate is thousands of times greater than the rate of extinction that prevailed before the appearance of human beings. Present-day extinctions, in which thousands of species are vanishing yearly, are forced extinctions caused by destructive human activities. If the spiral of human population growth and environmental destruction continues out of control, possibly by the middle of this century, half of all species living today will become extinct.

Already, many species nearing extinction can be found only in captivity. These represent only a small fraction of all endangered species. Once the population of a species drops below a certain critical level, the likelihood of extinction increases due to a greatly reduced gene pool. Inbreeding exposes species to genetic defects that decimate populations. Furthermore, once a species losses its genetic variability, it is doomed because the species can no longer adapt to changes in its environment.

In the past, climate changes were slow enough for the biologic world to adapt. However, today's climate changes are much too abrupt, perhaps rapid enough to cause the extinction of plants and animals. Plants would be hardest hit by global warming because they are directly affected by changes in temperature and rainfall. Forests, especially game preserves, might become isolated from their normal climate regimes, whose climatic conditions would continue to advance to higher latitudes. Human intervention on an unprecedented scale might be required to preserve plant and animal species threatened by climate change, especially if it occurs rapidly.

The effects of global climate change with increasing temperatures and rainfall would prevail for centuries. During that time, forests would creep poleward, while other wildlife habitats, including the Arctic tundra, would disappear. Many species would be unable to keep pace with these rapid climate changes. Those able to migrate could find their routes blocked by natural and artificial barriers, such as cities and farms. The warming would rearrange entire biologic communities and cause many species to become extinct, while others commonly called pests would overrun the landscape.

High levels of carbon dioxide, which functions as a fertilizer, favor the growth of weeds. The warmer climate would be a boon for parasites and pathogens, including bacteria and viruses, and could cause an influx of tropical diseases into the temperate zone. The culminating effect would be a diminishing species diversity worldwide. If the number of species continues to decline because of human interference with their habitats, our species might be threatened with extinction as well.

As a possible prelude to global extinction, amphibians, which have been in existence for more than 300 million years, are disappearing all over the world at an alarming rate. Amphibians such as frogs have developed deformities, including multiple or missing legs, possibly caused by pollutants such as

Figure 184 *Deforestation in the Tropics.*

(Courtesy National Center for Atmospheric Research)

pesticide and fertilizer runoff. As with all amphibians, frogs have permeable skins that can absorb toxins from the environment, making them extremely sensitive to environmental change. Since the 1960s, due to deforestation, acid rain, pollution, or ozone depletion, frog species have been going extinct in tragic numbers. Moreover, amphibians are vanishing from nature preserves, where little human perturbation occurs. These creatures might be sounding an early warning that the Earth is in grave danger.

Species are becoming extinct due to overhunting, introduced species that prey on or directly compete with indigenous species, destruction of habitats by deforestation (Fig. 184) along with other destructive human activities, and the collapse of food chains. Every species depends on other species for its survival. When too many species become extinct in an ecosystem, the remaining species are at risk of extinction by the "domino effect." If this process were to continue globally due to human destructive activities, it could initiate the collapse of biology and cause one of the worst extinction events in the history of our planet.

If trends continue, by the middle of this century, the number of extinct species could exceed those lost in the great extinctions of the geologic past. By our very actions, we are upsetting the delicate balance of nature. Should it tilt ever so slightly, cataclysmic changes could result. Mass extinctions normally occur over periods of thousands or even millions of years. However, because of our disturbance, the extinction of large numbers of species would take place in a mere century. We have yet to feel the adverse effects of such a huge die-off of species. However, once a wave of extinction is set in motion, it will ultimately undermine the quality or even the possibility of human life.

CONCLUSION

Volcanology has been called the science that marches forward only on the ashes of catastrophe. An erupting volcano is an obvious manifestation of the great heat engine that drives the Earth's plates around the surface of the globe. Volcanoes' role as a conduit for magma from the interior of the Earth is unmistakable. Yet their prediction still remains illusory. Although volcanoes can erupt almost anywhere, they are mostly concentrated at the junctions of crustal plates, where the Earth's geology is the most active. These regions are also the sites of most earthquakes. Along with vibrations from volcanoes, they produce some of the most shaky ground. Given the Earth's restless nature, additional volcanic crises will occur, especially in areas where human populations expand into areas with a high risk of volcanism.

Geologic hazards such as earthquakes and volcanic eruptions that have plagued civilization are merely slight readjustments in the Earth's crust. Earthquakes and volcanoes are caused by forces on the Earth that move huge eggshell-like plates over the surface of the planet. The plates form a brittle rock "skin" and are driven by the slow escape of the Earth's internal heat. Most earthquakes crack the Earth's crust like broken porcelain by a process of brittle fracturing and frictional sliding. The stress builds up until a fracture forms at shallow depths and slippage along the fault vibrates the ground. Catastrophe often results when this occurs near population centers.

Both earthquakes and volcanoes can be dangerous beyond their local domains. Coastal and undersea earthquakes can send powerful tsunamis racing across the ocean. Earthquakes also tend to loosen the ground on slopes, causing landslides that can be just as destructive as the temblors themselves. Volcanoes can spew large amounts of weather-altering materials into the atmosphere, causing global climate change. Volcano-induced mudflows and flooding can devastate vast areas. Volcanic clouds of ash and gas can reach out and incinerate or asphyxiate everything they touch. Volcanoes are also thought to initiate ice ages and cause the extinction of species. With these geologic hazards in mind, the Earth does not seem to be such a safe place to live in after all.

GLOSSARY

aa lava (AH-ah) a lava that forms large, jagged, irregular blocks
abrasion erosion by friction, generally caused by rock particles carried by running water, ice, and wind
agglomerate (ah-GLOM-eh-ret) a pyroclastic rock composed of consolidated volcanic fragments
albedo the amount of sunlight reflected from an object and dependent on its color and texture
alluvium (ah-LUE-vee-um) stream-deposited sediment
alpine glacier a mountain glacier or a glacier in a mountain valley
andesite an intermediate type of volcanic rock between basalt and rhyolite
anticline folded sediments that slope downward away from a central axis
Apollo asteroids asteroids that come from the main belt between Mars and Jupiter and cross the Earth's orbit
aquifer (AH-kwe-fer) a subterranean bed of sediments through which groundwater flows
arête (ah-RATE) a sharp-crested ridge formed by abutting cirques
ash fall the fallout of small, solid particles from a volcanic eruption cloud
asperite (AS-per-ite) the point where a fault hangs up and eventually slips, causing earthquakes
asteroid a rocky or metallic body whose impact on the Earth creates a large meteorite crater
asteroid belt a band of asteroids orbiting the Sun between the orbits of Mars and Jupiter

asthenosphere (as-THE-nah-sfir) a layer of the upper mantle from about 60 to 200 miles below the surface that is more plastic than the rock above and below and might be in convective motion

astrobleme (as-TRA-bleem) eroded remains on the Earth's surface of an ancient impact structure produced by a large cosmic body

avalanche (AH-vah-launch) a slide on a snowbank triggered by vibrations from earthquakes and storms

back-arc basin a seafloor-spreading system of volcanoes caused by extension behind an island arc that is above a subduction zone

barrier island a low, elongated coastal island that parallels the shoreline and protects the beach from storms

basalt (bah-SALT) a dark, volcanic rock that is usually quite fluid in the molten state

basement rock subterranean igneous, metamorphic, granitic, or highly deformed rock underlying younger sediments

batholith (BA-the-lith) the largest of intrusive igneous bodies and more than 40 square miles on its uppermost surface

bedrock solid layers of rock beneath younger materials

black smoker superheated hydrothermal water rising to the surface at a midocean ridge; the water is supersaturated with metals, and when exiting through the seafloor, it quickly cools and the dissolved metals precipitate, resulting in black, smokelike effluent

blowout a hollow caused by wind erosion

blue hole a water-filled sinkhole

bolide an exploding meteor whose fireball is often accompanied by a bright light and sound when passing through the earth's atmosphere

bomb, volcanic a solidified blob of molten rock ejected from a volcano

calcite a mineral composed of calcium carbonate

caldera (kal-DER-eh) a large, pitlike depression at the summits of some volcanoes and formed by great explosive activity and collapse

calving formation of icebergs by glaciers breaking off upon entering the ocean

carbonaceous (KAR-bah-NAY-shus) a substance containing carbon, namely sedimentary rocks such as limestone and certain types of meteorites

carbonate a mineral containing calcium carbonate such as limestone

carbon cycle the flow of carbon into the atmosphere and ocean, the conversion to carbonate rock, and the return to the atmosphere by volcanoes

catchment area the recharge area of a groundwater aquifer

Cenozoic (SIN-eh-zoe-ik) an era of geologic time comprising the last 65 million years

chondrite (KON-drite) the most common type of meteorite, composed mostly of rocky material with small, spherical grains

chondrule (KON-drule) rounded granules of olivine and pyroxine found in stony meteorites called chondrites

circum-Pacific active seismic regions around the rim of the belt Pacific plate coinciding with the Ring of Fire

cirque (serk) a glacial erosional feature, producing an amphitheater-like head of a glacial valley

col (call) a saddle-shaped mountain pass formed by two opposing cirques

coma the atmosphere surrounding a comet when it comes within the inner solar system; the gases and dust particles are blown outward by the solar wind to form the comet's tail

comet a celestial body believed to originate from a cloud of comets that surrounds the Sun and develops a long tail of gas and dust particles when traveling near the inner solar system

conduit a passageway leading from a reservoir of magma to the surface of the Earth through which volcanic products pass

cone, volcanic the general term applied to any volcanic mountain with a conical shape

continent a landmass composed of light, granitic rock that rides on the denser rocks of the upper mantle

continental glacier an ice sheet covering a portion of a continent

continental drift the concept that the continents have been drifting across the surface of the Earth throughout geologic time

continental margin the area between the shoreline and the abyss and that represents the true edge of a continent

continental shelf the offshore area of a continent in a shallow sea

continental shield ancient crustal rocks upon which the continents grew

continental slope the transition from the continental shelf to the deep-sea basin

convection a circular, vertical flow of a fluid medium by heating from below; as materials are heated, they become less dense and rise, cool, become more dense, and sink

convergent plate margin the boundary between crustal plates where the plates come together; generally corresponds to the deep-sea trenches where old crust is destroyed in subduction zones

coral any of a large group of shallow-water, bottom-dwelling marine invertebrates that build reef colonies in warm waters

Cordillera (KOR-dil-er-ah) a range of mountains that includes the Rockies, Cascades, and Sierra Nevada in North America and the Andes in South America

core the central part of the Earth, consisting of a heavy iron-nickel alloy

correlation (KOR-eh-LAY-shen) the tracing of equivalent rock exposures over distance usually with the aid of fossils

crater, meteoritic a depression in the crust produced by the bombardment of a meteorite

crater, volcanic the inverted conical depression found at the summit of most volcanoes, formed by the explosive emission of volcanic ejecta

craton (CRAY-ton) the stable interior of a continent, usually composed of the oldest rocks

creep the slow flowage of earth materials

crevasse (kri-VAS) a deep fissure in the crust or a glacier

crust the outer layers of a planet's or a moon's rocks

crustal plate a segment of the lithosphere involved in the interaction of other plates in tectonic activity

delta a wedge-shaped layer of sediments deposited at the mouth of a river

desertification (di-zer-te-fa-KA-shen) the process of becoming arid land

desiccated basin (de-si-KAY-ted) a basin formed when an ancient sea evaporated

diapir (DIE-ah-per) the buoyant rise of a molten rock through heavier rock

dike a tabular intrusive body that cuts across older strata

divergent margin the boundary between crustal plates where the plates move apart; generally corresponds to the midocean ridges where new crust is formed by the solidification of liquid rock rising from below

dolomite (DOE-leh-mite) a sedimentary rock formed by the replacement of calcium with magnesium in limestone

drumlin a hill of glacial debris facing in the direction of glacial movement

dune a ridge of windblown sediments usually in motion

earth flow the downslope movement of soil and rock

earthquake the sudden rupture of rocks along active faults in response to geologic forces within the Earth

East Pacific Rise a midocean-spreading center that runs north-south along the eastern side of the Pacific and the predominant location upon which the hot springs and black smokers have been discovered

elastic rebound theory the theory that earthquakes depend on rock elasticity

eolian (ee-OH-lee-an) a deposit of windblown sediment

epicenter the point on the Earth's surface directly above the focus of an earthquake

erosion the wearing away of surface materials by natural agents such as wind and water

erratic boulder a glacially deposited boulder far from its source

escarpment (es-KARP-ment) a mountain wall produced by the elevation of a block of land

esker (ES-ker) a curved ridge of glacially deposited material

evaporite (ee-VA-per-ite) the deposition of salt, anhydrite, and gypsum from evaporation in an enclosed basin of stranded seawater

exfoliation (eks-foe-lee-A-shen) the weathering of rock, causing the outer layers to flake off

extrusive (ik-STRU-siv) an igneous volcanic rock ejected onto the surface of a planet or moon

facies an assemblage of rock units deposited in a certain environment

fault a break in crustal rocks caused by earth movements

fissure a large crack in the crust through which magma might escape from a volcano

fjord (fee-ORD) a long, narrow, steep-sided inlet of a mountainous, glaciated coast

floodplain the land adjacent to a river that floods during river overflows

fluvial (FLUE-vee-al) pertaining to being deposited by a river

focus the point of origin of an earthquake; also called a hypocenter

formation a combination of rock units that can be traced over a distance

fossil any remains, impressions, or traces in rock of a plant or animal of a previous geologic age

frost heaving the lifting of rocks to the surface by the expansion of freezing water

frost polygons polygonal patterns of rocks from repeated freezing

fumarole (FUME-ah-role) a vent through which steam or other hot gases escape from underground such as a geyser

geomorphology (JEE-eh-more-FAH-leh-jee) the study of surface features of the Earth

geothermal the generation of hot water or steam by hot rocks in the Earth's interior

geyser (GUY-sir) a spring that ejects intermittent jets of steam and hot water

glacier a thick mass of moving ice occurring where winter snowfall exceeds summer melting

glacier burst a flood caused by an underglacier volcanic eruption

glacière (GLAY-sher-ee) an underground ice formation

Gondwana (gone-DWAN-ah) a southern supercontinent of Paleozoic time, comprising Africa, South America, India, Australia, and Antarctica, that broke up into the present continents during the Mesozoic era

graben (GRA-bin) a valley formed by a down-dropped fault block

granite a coarse-grained, silica-rich rock, consisting primarily of quartz and feldspars, and the principal constituent of the continents, believed to be derived from a molten state beneath the Earth's surface

gravity fault motion along a fault plane that moves as if pulled downslope by gravity; also called a normal fault

groundwater water derived from the atmosphere that percolates and circulates below the surface

guyot (GEE-oh) an undersea volcano that reached the surface of the ocean, whereupon its top was flattened by erosion; later, subsidence caused the volcano to sink below the surface

haboob (hey-BUBE) a violent dust storm or sandstorm

hanging valley a glaciated valley above the main glaciated valley often forming a waterfall

hiatus (hie-AY-tes) a break in geologic time due to a period of erosion or nondeposition of sedimentary rock

horn a peak on a mountain formed by glacial erosion

horst an elongated, uplifted block of crust bounded by faults

hot spot a volcanic center with no relation to a plate boundary; an anomalous magma generation site in the mantle

hyaloclastic (hi-AH-leh-KLAS-tic) basalt lava erupted beneath a glacier

hydrocarbon a molecule consisting of carbon chains with attached hydrogen atoms

hydrologic cycle the flow of water from the ocean to the land and back to the sea

hydrology the study of water flow over the Earth

hydrothermal relating to the movement of hot water through the crust; also a mineral ore deposit emplaced by hot groundwater

hypocenter the point of origin of earthquakes; also called focus

Iapetus Sea (EYE-ap-i-tus) a former sea that occupied a similar area as the present Atlantic Ocean prior to Pangaea

ice age a period of time when large areas of the Earth were covered by massive glaciers

iceberg a portion of a glacier calved off upon entering the sea

ice cap a polar cover of snow and ice

igneous rocks all rocks solidified from a molten state

ignimbrite (IG-nem-brite) a hard rock composed of consolidated pyroclastic material

impact the point on the surface upon which a celestial object has landed, creating a crater

interglacial a warming period between glacial periods

intertidal zone the shore area between low and high tides

intrusive any igneous body that has solidified in place below the Earth's surface

iridium (i-RI-dee-em) a rare isotope of platinum, relatively abundant on meteorites

island arc volcanoes landward of a subduction zone, parallel to a trench, and above the melting zone of a subducting plate

isostasy (eye-SOS-the-see) a geologic principle that states that the Earth's crust is buoyant and rises and sinks depending on its density

isotope (I-seh-tope) a particular atom of an element that has the same number of electrons and protons as the other atoms of the element but a different number of neutrons; i.e., the atomic numbers are the same, but the atomic weights differ

jointing the production of parallel fractures in rock formations

kame a steep-sided mound of moraine deposited at the margin of a melting glacier

karst a terrain comprised of numerous sinkholes in limestone

kettle a depression in the ground caused by a buried block of glacial ice

kimberlite (KIM-ber-lite) a volcanic rock composed mostly of peridotite, originating deep within the mantle and that brings diamonds to the surface

Kirkwood gaps bands in the asteroid belt that are mostly empty of asteroids due to Jupiter's gravitational attraction

lahar (LAH-har) a mudflow of volcanic material on the flanks of a volcano

lamellae (leh-ME-lee) striations on the surface of crystals caused by a sudden release of high pressures such as those created by large meteorite impacts

landslide a rapid downhill movement of earth materials triggered by earthquakes and severe weather

lapilli (leh-PI-lie) small, solid pyroclastic fragments

lateral moraine the material deposited by a glacier along its sides

Laurasia (LURE-ay-zha) a northern supercontinent of Paleozoic time consisting of North America, Europe, and Asia

Laurentia (LURE-in-tia) an ancient North American continent

lava molten magma that flows out onto the surface

limestone a sedimentary rock consisting mostly of calcite from shells of marine invertebrates

liquefaction (li-kwe-FAK-shen) the loss of support of sediments that liquefy during an earthquake

lithosphere (LI-the-sfir) the rocky outer layer of the mantle that includes the terrestrial and oceanic crusts; the lithosphere circulates between the Earth's surface and mantle by convection currents

lithospheric plate a segment of the lithosphere involved in the plate interaction of other plates during tectonic activity

loess (LOW-es) a thick deposit of airborne dust

magma a molten rock material generated within the Earth that is the constituent of igneous rocks

magnetic field reversal a reversal of the north-south polarity of the magnetic poles

magnetometer a device used to measure the intensity and direction of the magnetic field

magnitude scale a scale for rating earthquake energy

mantle the part of a planet below the crust and above the core, composed of dense rocks that might be in convective flow

mass wasting the downslope movement of rock under the direct influence of gravity

Mesozoic (MEH-zeh-ZOE-ik) literally the period of middle life, referring to a period between 250 and 65 million years ago

metamorphism (me-teh-MORE-fi-zem) recrystallization of previous igneous, metamorphic, or sedimentary rocks created under conditions of intense temperatures and pressures without melting

meteor a small, celestial body that becomes visible as a streak of light when entering the Earth's atmosphere

meteorite a metallic or stony celestial body that enters the Earth's atmosphere and impacts onto the surface

meteoritics the science that deals with meteors and related phenomena

meteoroid a meteor in orbit around the Sun with no relation to the phenomenon it produces when entering the Earth's atmosphere

meteor shower a phenomenon observed when large numbers of meteors enter the Earth's atmosphere; the meteors' luminous paths appear to diverge from a single point

microearthquake a small earth tremor

micrometeorites small, grain-sized bodies that strike spacecraft

microtektites small, spherical grains created by the melting of surface rocks during a large meteorite impact

Mid-Atlantic Ridge the seafloor-spreading ridge that marks the extensional edge of the North and South American plates to the west and the Eurasian and African plates to the east

midocean ridge a submarine ridge along a divergent plate boundary where a new ocean floor is created by the upwelling of mantle material

moraine (mah-RANE) a ridge of erosional debris deposited by the melting margin of a glacier

moulin (mue-LIN) a cylindrical shaft extending down into a glacier and produced by meltwater

mountain roots the deeper crustal layers under mountains

mudflow the flowage of sediment-laden water

nonconformity an unconformity in which sedimentary deposits overlie crystalline rocks

normal fault a gravity fault in which one block of crust slides down another block of crust along a steeply tilted plane thrust onto continents by plate collisions.

nuée ardente (NU-ee ARE-dent) a volcanic pyroclastic eruption of hot ash and gas

Oort Cloud the collection of comets that surround the Sun about a light-year away

ophiolite (OH-fi-ah-lite) masses of oceanic crust thrust onto the continents by plate collisions

orogen (ORE-ah-gin) an eroded root of an ancient mountain range

orogeny (oh-RAH-ja-nee) a process of mountain building by tectonic activity

outgassing the loss of gas from within a planet as opposed to degassing, the loss of gas from meteorites

overthrust a thrust fault in which one segment of crust overrides another segment for a great distance

pahoehoe lava (pah-HOE-ay-hoe-ay) a lava that forms ropelike structures when cooled

paleomagnetism the study of the Earth's magnetic field, including the position and polarity of the poles in the past

paleontology (PAY-lee-ON-tah-logy) the study of ancient life-forms, based on the fossil record of plants and animals

Paleozoic (PAY-lee-eh-ZOE-ik) the period of ancient time between 570 and 250 million years ago

Pangaea (PAN-gee-a) an ancient supercontinent that included all the lands of the Earth

Panthalassa (PAN-the-lass-a) a great global ocean that surrounded Pangaea

peridotite (pah-RI-deh-tite) the most common rock type in the mantle

periglacial referring to geologic processes at work adjacent to a glacier

permafrost permanently frozen ground in the Arctic regions

permeability the ability to transfer fluid through cracks, pores, and interconnected spaces within a rock

pillow lava lava extruded onto the ocean floor, giving rise to tabular shapes

placer (PLAY-ser) a deposit of rocks left behind by a melting glacier; any ore deposit enriched by stream action

planetoid a small body, generally no larger than the Moon, in orbit around the Sun; a disintegration of several such bodies might have been responsible for the asteroid belt between Mars and Jupiter

plate tectonics the theory that accounts for the major features of the Earth's surface in terms of the interaction of lithospheric plates

playa (PLY-ah) a flat, dry, barren plain at the bottom of a desert basin

pluton (PLUE-ton) an underground body of igneous rock younger than the rocks that surround it and that formed where molten rock oozed into a space between older rocks

pumice volcanic ejecta that is full of gas cavities and extremely light in weight

pyroclastic (PIE-row-KLAS-tik) the fragmental ejecta released explosively from a volcanic vent

radiometric dating determining the age of an object by chemically analyzing its stable and unstable radioactive elements

recessional moraine a glacial moraine deposited by a retreating glacier

reef the biological community that lives at the edge of an island or continent; the shells from dead organisms form a limestone deposit

regression a fall in sea level, exposing continental shelves to erosion

resurgent caldera a large caldera that experiences renewed volcanic activity and that domes up the caldera floor

rhyolite (RYE-oh-lite) a volcanic rock that is highly viscous in the molten state and usually ejected explosively as pyroclastics

rift valley the center of an extensional spreading where continental or oceanic plate separation occurs

rille (ril) a trench formed by a collapsed lava tunnel

riverine (RI-vah-rene) relating to a river

roche moutonnée (ROSH mue-tin-ay) a knobby, glaciated bedrock surface

saltation the movement of sand grains by wind or water

salt dome an upwelling plug of salt that arches surface sediments and often serves as an oil trap

sand boil an artesian-like fountain of sediment-laden water produced by the liquefaction process during an earthquake

scarp a steep slope formed by earth movements

seafloor spreading a theory that the ocean floor is created by the separation of lithospheric plates along midocean ridges with new oceanic crust formed from mantle material that rises from the mantle to fill the rift

seamount a submarine volcano

seiche (seech) a wave oscillation on the surface of a lake or landlocked sea

seismic (SIZE-mik) pertaining to earthquake energy or other violent ground vibrations

seismic sea wave an ocean wave generated by an undersea earthquake or volcano; also called tsunami

seismometer a detector of earthquake waves

shield areas of exposed Precambrian nucleus of a continent

shield volcano a broad, low-lying volcanic cone built up by lava flows of low viscosity

sinkhole a large pit formed by the collapse of surface materials undercut by the solution of subterranean limestone

solifluction (SOE-leh-flek-shen) the failure of earth materials in tundra

spherules (SFIR-ule) small, spherical, glassy grains found on certain types of meteorites, on lunar soils, and at large meteorite impact sites on Earth

stishovite (STIS-hoe-vite) a quartz mineral produced by extremely high pressures such as those generated by a large meteorite impact

stratovolcano an intermediate volcano characterized by a stratified structure from alternating emissions of lava and fragments

strewn field a usually large area where tektites are found arising from a large meteorite impact

striae (STRY-aye) scratches on bedrock made by rocks embedded in a moving glacier

subduction zone a region where an oceanic plate dives below a continental plate into the mantle; ocean trenches are the surface expression of a subduction zone

subsidence the compaction of sediments due to the removal of fluids

surge glacier a continental glacier that heads toward the sea at a high rate of advance

syncline (SIN-kline) a fold in which the beds slope inward toward a common axis

talus cone a steep-sided pile of rock fragments at the foot of a cliff

tarn a small lake formed in a cirque

tectonics (tek-TAH-nik) the history of the Earth's larger features (rock formations and plates) and the forces and movements that produce them

tektites (TEK-tite) small, glassy minerals created from the melting of surface rocks by the impact of a large meteorite

tephra (THE-fra) all clastic materials from dust particles to large chunks, expelled from volcanoes during eruptions

terrane (teh-RAIN) a unique crustal segment attached to a landmass

Tethys Sea (THE-this) the hypothetical midlatitude region of the oceans separating the northern and southern continents of Laurasia and Gondwana

till sedimentary material deposited by a glacier

tillite a sedimentary deposit composed of glacial till

transform fault a fracture in the Earth's crust along which lateral movement occurs; they are common features of the midocean ridges

transgression a rise in sea level that causes flooding of the shallow edges of continental margins

trapps (traps) a series of massive lava flows that resembles a staircase

trench a depression on the ocean floor caused by plate subduction

tsunami (sue-NAH-me) a seismic sea wave produce by an undersea or nearshore earthquake or volcanic eruption

tuff a rock formed of pyroclastic fragments

tundra permanently frozen ground at high latitudes and elevations
varves thinly laminated lake bed sediments deposited by glacial meltwater
ventifact (VEN-te-fakt) a stone shaped by the action of windblown sand
verga precipitation that evaporates before reaching the ground
volcanic ash fine pyroclastic material injected into the atmosphere by an erupting volcano
volcanic bomb a solidified blob of molten rock ejected from a volcano
volcano a fissure or vent in the crust through which molten rock rises to the surface to form a mountain
wadi or wash the dry bed of a stream that flows with water during a heavy downpour in a desert

BIBLIOGRAPHY

THE DYNAMIC EARTH

Allegre, Claud J. and Stephen H. Snider. "The Evolution of the Earth." *Scientific American* 271 (October 1994): 66–75.

Dickinson, William R. "Making Composite Continents." *Nature* 364 (July 22, 1993): 284–285.

Gordon, Richard G. and Seth Stein. "Global Tectonics and Space Geodesy." *Science* 256 (April 17, 1992): 333–341.

Jeanloz, Raymond and Thorne Lay. "The Core-Mantle Boundary." *Scientific American* 268 (May 1993): 48–55.

Kunzig, Robert. "Birth of a Nation." *Discover* 11 (February 1990): 26–27.

Macdonald, Kenneth C. and Paul J. Fox. "The Mid-Ocean Ridge." *Scientific American* 262 (June 1990): 72–79.

Murphy, J. Brendan and R. Damian Nance. "Mountain Belts and the Supercontinent Cycle." *Scientific American* 266 (April 1992): 84–91.

Peacock, Simon M. "Fluid Processes in Subduction Zones." *Science* 248 (April 20, 1990): 329–336.

Shurkin, Joel and Tom Yulsman. "Assembling Asia." *Earth* 4 (June 1995): 52–59.

Taylor, S. Ross and Scott M. McLennan. "The Evolution of Continental Crust." *Scientific American* 274 (January 1996): 76–81.

Weiss, Peter. "Land Before Time." *Earth* 8 (February 1998): 29–33.

Zimmer, Carl. "Ancient Continent Opens Window on the Early Earth." *Science* 286 (December 17, 1999): 2254–2256.

EARTHQUAKES

Bolt, Bruce A. "Balance of Risks and Benefits in Preparation for Earthquakes." *Science* 251 (January 11, 1991): 169–174.
Fischman, Joshua. "Falling Into the Gap." *Discover* 13 (October 1992): 57–63.
Folger, Tim. "Waves of Destruction." *Discover* 15 (May 1994): 68–73.
Frohlich, Cliff. "Deep Earthquakes." *Scientific American* 260 (January 1989): 48–55.
Gonzalez, Frank I. "Tsunami!" *Scientific American* 280 (May 1999): 56–65.
Green, Harry W., II. "Solving the Paradox of Deep Earthquakes." *Scientific American* 271 (September 1994): 64–71.
Heaton, Thomas H. and Stephen H. Hartzell. "Earthquake Hazards on the Cascadia Subduction Zone." *Science* 236 (April 10, 1987) 162–168.
Johnston, Arch C. and Lisa R. Kanter. "Earthquakes in Stable Continental Crust." *Scientific American* 262 (March 1990): 68–75.
Kerr, Richard A. "What Makes the San Andreas Tick?" *Science* 254 (October 11, 1991): 197–198.
Normile, Dennis. "A Wake-up Call From Kobe." *Popular Science* (February 1996): 64–68.
Pendick, Daniel. "Himalayan High Tension." *Earth* 5 (October 1996): 46–53.
Roman, Mark B. "Finding Fault." *Discover* 9 (August 1988): 57–63.
Stein, Ross S. and Robert S. Yeats. "Hidden Earthquakes." *Scientific American* 260 (June 1989): 48–57.
Unklesbay, A. G. "Midwest Earthquakes." *Earth Science* 40 (Winter 1987): 11–13.
Vogel, Shawna. "Shocks Heard Round the World." *Discover* 11 (January 1990): 68–70.

VOLCANIC ERUPTIONS

Berreby, David. "Barry Versus the Volcano." *Discover* 12 (June 1991): 60–67.
Chen, Allan. "The Thera Theory." *Discover* 10 (February 1989): 77–83.
Dvorak, John J., Carl Johnson, and Robert I. Tilling. "Dynamics of Kilauea Volcano." *Scientific American* 267 (August 1992): 46–53.
Kerr, Richard A. "Volcanoes: Old, New, and-Perhaps-Yet to Be." *Science* 250 (December 21, 1990): 1660–1661.
Krajick, Kevin. "To Hell and Back." *Discover* 20 (July 1999): 76–82.

Lewis, G. Brad. "Island of Fire." *Earth* 4 (October 1995): 32–33.
Lipman, Peter. "Chasing the Volcano." *Earth* 6 (December 1997): 32–39.
Lockridge, Patricia A. "Volcanoes and Tsunamis." *Earth Science* 42 (Spring 1989): 24–25.
Milstein, Michael. "Cooking up a Volcano." *Earth* 7 (April 1998): 24–30.
Mouginis-Mark, Peter, J. "Volcanic Hazards Revealed by Radar Interferometry." *Geotimes* 39 (July 1994): 11–13.
Newhall, Christopher G. "Mount St. Helens, Master Teacher." *Science* 288 (May 19, 2000): 68–80.
Roche, Kirby. "The Mystique of Disaster." *Weatherwise* 43 (October 1990): 262–264.
Stager, Curt. "Africa's Great Rift." *National Geographic* 177 (May 1990): 10–41.
Wickelgren, Ingrid. "Simmering Planet." *Discover* 11 (July 1990): 73–75.

EARTH MOVEMENTS

Finkbeiner, Ann. "Terra Infirma." *Discover* 12 (November 1991): 18.
Friedman, Gerald M. "Slides and Slumps." *Earth Science* 41 (Fall 1988): 21–23.
Glanz, James. "Erosion Study Finds High Price for Forgotten Menace." *Science* 267 (February 24, 1995): 1088.
Kerr, Richard A. "Volcanoes With Bad Hearts Are Tumbling Down All Over." *Science* 264 (April 29, 1994): 660.
Monastersky, Richard. "When Kilauea Crumbles." *Science News* 147 (April 8, 1995): 216–218.
———. "When Mountains Fall." *Science News* 142 (August 29, 1992): 136–138.
———. "Soil May Signal Imminent Landslide." *Science* News 134 (November 12, 1988): 318.
Norris, Robert M. "Sea Cliff Erosion." *Geotimes* 35 (November 1990): 16–17.
Pendick, Daniel. "Ashes, Ashes, All Fall Down." *Earth* 6 (February 1997): 32–33.
Peterson, Ivars. "Digging into Sand." *Science News* 136 (July 15, 1989): 40–42.
Shaefer, Stephen J. and Stanley N. Williams. "Landslide Hazards." *Geotimes* 36 (May 1991): 20–22.
Simpson, Sarah, "Raging Rivers of Rock." *Scientific American* 283 (July 2000): 24–25.
Zimmer, Carl. "Landslide Victory." *Discover* 12 (February 1991): 66–69.

CATASTROPHIC COLLAPSE

Bolton, David W. "Underground Frontiers." *Earth Science* 40 (Summer 1987): 16–18.

Francis, Peter. "Giant Volcanic Calderas." *Scientific American* 248 (June 1983): 60–70.

Francis, Peter and Stephen Self. "Collapsing Volcanoes." *Scientific American* 256 (June 1987): 91–97.

Holzer, T. L., T. L. Youd, and T. C. Hanks. "Dynamics of Liquefaction During the 1987 Superstition Hills, California Earthquake." *Science* 244 (April 7, 1989): 56–59.

Hon, Ken and John Pallister. "Wrestling with Restless Calderas and Fighting Floods of Lava." *Nature* 376 (August 17, 1995): 554–555.

Lipske, Mike. "Wonder Holes." *International Wildlife* 20 (February 90): 47–51.

Meyer, Alfred. "Between Venice & the Deep Blue Sea." *Science 86* 7 (July/August 1986): 50–57.

Monastersky, Richard. "Against the Tide." *Science News* 156 (July 24, 1999): 63.

Nuhfer, Edward B. "What's a Geologic Hazard?" *Geotimes* 39 (July 1994): 4.

Pollack, Henry N. and David S. Chapman. "Underground Records of Changing Climate." *Scientific American* 268 (June 1993): 44–50.

Simon, Cheryl. "A Giant's Troubled Sleep." *Science News* 124 (July 16, 1983): 40–41.

Svitil, Kathy A. "A Giant's Malaise." *Discover* 15 (March 1995): 26.

Weisburd, Stefi. "Sensing the Voids Underground." *Science News* 130 (November 22, 1986): 329.

FLOODS

Abelson, Philip H. "Climate and Water." *Science* 243 (January 27, 1989): 461.

Adler, Jerry. "Troubled Waters." *Newsweek* (June 26, 1993): 21–27.

Cathles, Lawrence M., III. "Scales and Effects of Fluid Flow in the Upper Crust." *Science* 248 (April 20, 1990): 323–328.

Folger, Tim. "Waves of Destruction." *Discover* 15 (May 1994): 68–73.

LeComte, Douglas. "Wide Extremes and Drought." *Weatherwise* 30 (February 1984): 9–17.

Macilwain, Colin. "Conservationists Fear Defeat on Revised Flood Control Policies." *Nature* 365 (October 7, 1993): 468.

McCormick, John. "Washed Away." *Newsweek* (May 5, 1997): 29–31.

Monastersky, Richard. "Volcanoes Under Ice: Recipe for a Flood." *Science News* 150 (November 23, 1996): 327.

———. "Rivers in a Greenhouse World." *Science News* 137 (June 9, 1990): 365.

Nichols, Frederic H., et al. "The Modification of an Estuary." *Science* 231 (February 7, 1986): 567–573.

Ramage, Colin S. "El Niño." *Scientific American* 254 (June 1983): 77–83.

Webster, Peter J. "Monsoons." *Scientific American* 245 (August 1981): 109–118.

Williams, Nigel. "Dams Drain the Life Out of Riverbanks." *Science* 276 (May 2, 1997): 683.

DUST STORMS

Broeker, Wallace S. "Chaotic Climate." *Scientific American* 262 (November 1995): 62–68.
Campbell, John M. "Desert Extremes." *Earth* 7 (April 1998): 39–46.
D'Agnese Joseph. "Why Has Our Weather Gone Wild?" *Discover* 21 (June 2000): 72–78.
Gregory, Jonathan. "The Coming Climate." *Scientific American* 276 (May 1997): 79–83.
Hedin, Lars O. and Gene E. Likens. "Atmospheric Dust and Acid Rain." *Scientific American* 275 (December 1996): 88–92.
Mack, Walter N. and Elizabeth A. Leistikow. "Sands of the World." *Scientific American* 275 (August 1996): 62–67.
Nori, Franco, et al. "Booming Sand." *Scientific American* 277 (September 1997): 84–89.
Pennisi, Elizabeth. "Dancing Dust." *Science News* 142 (October 3, 1992): 218–220.
Raloff, Janet. "Holding on to the Earth." *Science News* 144 (October 1993): 280–281.
Reganold, John P., Robert I. Papendick, and James F. Parr. "Sustainable Agriculture." *Scientific American* 262 (June 1990): 112–120.
Repetto, Robert. "Deforestation in the Tropics." *Scientific American* 262 (April 1990): 36–42.
White, Robert M. "The Great Climate Debate." *Scientific American* 263 (July 1990): 36–43.
Zimmer, Carl. "How to Make a Desert." *Discover* 16 (February 1995): 51–56.

GLACIERS

Broecker, Wallace S. and George H. Denton. "What Drives Glacial Cycles?" *Scientific American* 262 (January 1990): 49–56.
Clark, Peter U. "Fast Glacier Flow Over Soft Beds." *Science* 267 (January 6, 1995): 43–44.
Gordon, Arnold L. and Josefino C. Comiso. "Polynyas in the Southern Ocean." *Scientific American* 258 (June 1988): 90–97.
Hoffman, Paul F. and David P. Schrag. "Snowball Earth." *Scientific American* 282 (January 2000): 68–75.

Horgan, John. "The Big Thaw." *Scientific American* 274 (November 1995): 18–20.
———. "Antarctic Meltdown." *Scientific American* 268 (March 1993): 19–28.
Kimber, Robert. "A Glacier's Gift." *Audubon* 95 (May–June 1993): 52–53.
Maslin, Mark. "Waiting for the Polar Meltdown." *New Scientist* 139 (September 4, 1993): 36–41.
Mathews, Samuel W. "Ice on the World." *National Geographic* 171 (January 1987): 84–103.
Mollenhauer, Erik and George Bartunek. "Glaciers on the Move." *Earth Science* 41 (Spring 1988): 21–24.
Moran, Joseph M., Ronald D. Stieglitz, and Donn P. Quigley. "Glacial Geology." *Earth Science* 41 (Winter 1988): 16–18.
Parfit, Michael. "Timeless Valleys of the Antarctic Desert." *National Geographic* 194 (October 1998): 120–135.
———. "Antarctic Meltdown." *Discover* 10 (September 1989): 39–47.
Peltier, Richard W. "Ice Age Paleotopography." *Science* 265 (July 8, 1994): 195–201.
Waters, Tom. "A Glacier Was Here." *Earth* 4 (February 1995): 58–60.

IMPACT CRATERING

Alvarez, Walter and Frank Asaro. "An Extraterrestrial Impact." *Scientific American* 263 (October 1990): 78–84.
Binzel, Richard P., M. Antonietta, and Marcello Fulchignoni. "The Origins of the Asteroids." *Scientific American* 265 (October 1991): 88–94.
Desonie, Dana. "The Threat from Space." *Earth* 5 (August 1996): 25–31.
Gehrels, Tom. "Collisions with Comets and Asteroids." *Scientific American* 274 (March 1996): 54–59.
Grieve, Richard A. F. "Impact Cratering on the Earth." *Scientific American* 262 (April 1990): 66–73.
Hildebrand, Alan R. and William V. Boynton. "Cretaceous Ground Zero." *Natural History* (June 1991): 47–52.
Long, Michael E. "Mars on Earth." *National Geographic* 196 (July 1999): 34–51.
Mathews, Robert. "A Rocky Watch for Earthbound Asteroids." *Science* 255 (March 6, 1992): 1204–1205.
Monastersky, Richard. "Target Earth." *Science News* 153 (May 16, 1998): 312–314.
———. "Shots from Outer Space." *Science News* 147 (January 28, 1995): 58–59.
Morrison, David. "Target Earth: It Will Happen." *Sky & Telescope* 79 (March 1990): 261–265.
Roach, Mary. "Meteorite Hunters." *Discover* 18 (May 1997): 71–75.
Sharpton, Virgil L. "Glasses Sharpen Impact Views." *Geotimes* 33 (June 1988): 10–11.

Stone, Richard. "The Last Great Impact on Earth." *Discover* 17 (September 1996): 60–71.

Weissman, Paul R. "Are Periodic Bombardments Real?" *Sky & Telescope* 79 (March 1990): 266–270.

MASS EXTINCTIONS

Courtillot, Vincent E. "A Volcanic Eruption." *Scientific American* 263 (October 1990): 85–92.

Crowley, Thomas J. and Gerald R. North. "Abrupt Climate Change and Extinction Events in Earth History." *Science* 240 (May 20, 1988): 996–1001.

Diamond, Jared. "Playing Dice with Megadeath." *Discover* 11 (April 1990): 55–59.

Ehrlich, Paul R. and Edward O. Wilson. "Biodiversity Studies: Science and Policy." *Science* 253 (August 16, 1991): 758–761.

Eldredge, Niles. "What Drives Evolution?" *Earth* 5 (December 1996): 34–37.

Erwin, Douglas H. "The Mother of Mass Extinctions." *Scientific American* 275 (July 1996): 72–78.

Flannery, Tim. "Debating Extinction." *Science* 283 (January 8, 1999): 182–183.

Gould, Stephen Jay. "The Evolution of Life on the Earth." *Scientific American* 271 (October 1994): 85–91.

King, Michael D. and David D. Herring. "Monitoring Earth's Vital Signs." *Scientific American* 282 (April 2000): 92–97.

Levington, Jeffrey S. "The Big Bang of Animal Evolution." *Scientific American* 267 (November 1992): 84–91.

May, Robert M. "How Many Species Inhabit the Earth?" *Scientific American* 267 (October 1992): 42–48.

Monastersky, Richard. "The Rise of Life on Earth." *National Geographic* 194 (March 1998): 54–81.

Orgel, Leslie E. "The Origin of Life on the Earth." *Scientific American* 271 (October 1994): 77–83.

Schmidt, Karen. "Life on the Brink." *Earth* 6 (April 1997): 26–33.

Simpson, Sarah, "Looking for Life Below." *Scientific American* 282 (June 2000): 94–101.

Terborgh, John. "Why American Songbirds Are Vanishing." *Scientific American* 266 (May 1992): 98–104.

Wilson, Edward O. "Threats to Biodiversity." *Scientific American* 261 (September 1989): 108–116.

INDEX

Boldface page numbers indicate extensive treatment of a topic. *Italic* page numbers indicate illustrations or captions. Page numbers followed by *m* indicate maps; *t* indicate tables; *g* indicate glossary.

D

E

F

G

H

I

J

K

L

M

N

O

P

Q

R

T

U

V

W

Y

Z